SOIL AND CROP PRODUCTIVITY

SOIL & CROP PRODUCTIVITY

A Handbook for the use of Workers in Agricultural Extension, Vocational Agriculture and Farmers in India

S. V. Govinda Rajan
H. G. Gopala Rao

ASIA PUBLISHING HOUSE
LONDON

Salem Venkatappa **Govinda Rajan** (1913)
Hunsur Gururao **Gopala Rao** (1913)

ISBN 0 210 27100 0

PRINTED IN INDIA
AT NATIONAL PRINTING WORKS, 10 DARYAGANJ, DELHI-6 AND PUBLISHED BY P.S. JAYASINGHE, ASIA PUBLISHING HOUSE, 447 STRAND, LONDON W.C. 2

PREFACE

The purpose of this book is to consider the various aspects of soils and their properties in relation to crop production. An attempt is made to present the salient properties of soils—both in respect of their physical formation and make-up and also the factors that play important roles in influencing crop production—in a manner that would be understood by students and workers in the field of agriculture. While aspects of a fundamental nature are presented in a manner that can be considered suitable for use by students of agriculture at the undergraduate levels in the colleges, considerable attention is paid to the applied aspects of soils and their use under Indian conditions which can be useful to workers in the field.

After a brief and direct presentation of the factors that can be considered important in deciding the qualities of a good productive soil, as compared with those of soils which do not possess these attributes, consideration is given to the formation of soils and the methods by which useful properties in them, like texture and structure, can be recognised. A chapter is devoted to a consideration of the occurrence and formation of different soils in India and their characteristics. The recognition of soil features and characters through field studies and the drawing up of an inventory of soils that occur in specified areas through surveys, have assumed considerable importance. A chapter is devoted to the practical aspects of such characterization of soils through reconnaissance and detailed surveys and the mapping of soils. Soil characterization has practical use in helping the drawing up of optimum land use classes. Rational utilization of soil resources is important from the point of obtaining the maximum benefit through sustained production. Land use classification and the manner of interpretation of the soil data for different management practices are dealt with in a manner which can be easily understood by both students of agriculture and followed by field workers in extension and vocational agriculture.

The different aspects of fertilizer use, including the use of lime, are dealt with in the perspective of practical soil tests. Soil conservation besides providing for anti-erosion measures, in the broadest sense, includes the adoption of all measures aimed at maintaining the soil conditions and fertility of the soil at optimum levels and these aspects are covered in detail in successive chapters. The important aspects of irrigation and drainage as also the methods of recognition of the factors of salinity and the management of salinity and alkalinity affected lands are dealt with in a manner that will be useful for application in the field. Illustrative material has been chosen from some of the sources as are available from work in our country. There is considerable experimental work published in our country in support of the various conclusions drawn on the different subjects, but, beyond a few essential references, it has not been thought necessary to burden this book with numerous references.

The authors have attempted to include in this text their experiences gained in teaching the subject matter for several years in the Mysore Agricultural College. Further, in assembling the material for this book, the works of numerous authors and scientists in the country has been drawn upon. Many illustrations are included with a view to bringing out prominently the subjects discussed and while these may not be entirely comprehensive, it is our belief that these will be very useful to the field workers. Special acknowledgements are made to Dr. Roy L. Donahue, Soils Consultant, Ford Foundation, who, besides furnishing numerous photographs to be included in this book, has given useful suggestions on the development of the subject matter incorporated in the book. Acknowledgements are also made to Mr. P.R. Raghuram Pillai, Artist of the Department of Agriculture, Bangalore, for furnishing several photographic illustrations.

The chapter on land use classification contains considerable information on interpretation of soil data for such classification based on the practical work in the field under the All India Soil and Land Use Survey and also discussions with Dr. N.R. Datta Biswas, Soil Conservation Officer, acknowledgements for which are made here. With regard to the preparation of the Ms for Press, particular mention is made of the help given by Mr. T.R. Tandon and Mr. Parminder Singh. The assistance given in the preparation of certain sketches and drawings by the staff of the

Cartographic Laboratory is acknowledged.

In recent years the cultivation of high yielding varieties of wheat, maize, paddy, sorghum and others have made great strides and have been popularised in all parts of the country. The higher intake of plant nutrients by these crops are met by application of larger and balanced mixtures of N, P and K, and also essential minor elements. Considerable work has been carried out to study the fertilizer responses of different varieties of these high yielding crops and it has been found opportune to include in this book data on some of the important trials conducted in this direction, and also to detail the current recommendations of fertilizer dosages for these crops as suited to conditions in different parts of the country. To Dr. M.S. Swaminathan, Director, I.A.R.I., we are grateful for his suggestion to include data on these, which help to make the book more useful to the reader.

Grateful thanks are also due to Dr. A.B. Joshi, Former Director, I.A.R.I., for his encouragement and help.

All India Soil and Land Use Survey,
I. A. R. I.,
New Delhi

S.V. Govinda Rajan
H.G. Gopala Rao

CONTENTS

CHAPTER I

SOIL AND THE CROP

WHEN a farmer plans to raise a crop on his field, almost the first thing that he considers is the nature of the soil present on the land. A variety of questions arises in his mind as regards the prospects of successful cropping. Is the soil fertile enough to raise a successful crop? Can it bear a heavy crop? Is it deep enough so that the plant roots can grow and develop satisfactorily? Does the soil possess the properties for storing the moisture supplied through rainfall or irrigation so that the crop is raised with a minimum of labour and cost? Are there any harmful conditions present which may affect the normal growth of the crop? These are some of the questions that will arise in his mind. There are many other questions also which will assail his mind and most of these will require a knowledge of the qualities and properties of soils and their relationship to successful crop production. A good working knowledge of the nature of the soils of his fields and their relationship to the environment is, therefore, necessary to give him enough confidence to go ahead with his plans and this will help him to have sound judgement in regard to the various operations that he will have to carry out in cultivating his lands and in raising his crops with success.

SOIL : AN ESSENTIAL MEDIUM FOR RAISING CROPS

It is worth while for all those engaged in agriculture to know what role the soil plays in crop production and what are the various management practices needed to maintain the soils in proper physical condition and for sustained crop production. First, it is a truism to state that soil bears the crops. The land and the soil that the farmer tills are prerequisites for raising a crop. Soil is the medium in which the crop has to establish the roots of the young plant, permit its growth, and spread its roots safely anchored within the surface of the land, so that the plant can complete its growth cycle. Attempts have been made in modern times to raise crops by avoiding the use of soil, through the techniques of water culture, termed *Hydroponics*, using nutrient solutions only. But these techniques have been tried only on a limited scale and they are uneconomical for field-scale operations except under special conditions and for particular crops. By and large, the world's population has to depend upon soil of the earth's crust to bear the crops that require to be grown for the needs of human and animal life. Further, soil is a natural source of supply of all the essential mineral nutrients required for plant growth, some nutrients, like carbon, hydrogen, and oxygen being obtained from the atmosphere and water. It is, therefore, to be expected that the exploitation of the resources of land and the soil will continue to be the only method for raising crops needed by the human population. The continued exploitation of the soil will cause diminution or depletion from the

Fig. 1. A good stand of paddy crop. Proper care in selection of seeds, adequate water supply, manures and fertilizers besides care and management of the soil ensure high crop production.

soil of the important constituents needed by plants for their nutrition and growth, and these will need to be made good by the application of suitable manures and fertilizers, so that their cropping capacity is maintained at a satisfactory level. As such, a knowledge of the qualities and the properties of the medium of the soil, which is so essential for economical crop production, is an important prerequisite for planning successful cropping.

FERTILE SOIL AND PRODUCTIVE SOIL

At this stage we may usefully distinguish between soil fertility and soil productivity. A fertile soil may be defined as one which has a good supply of available plant nutrients to be drawn upon by plants throughout their growth period. A productive soil has the capacity to produce good crops, and this capacity is derived not

FIG. 2. A luxuriant growth of tobacco on a productive soil. Proper attention to cultivation practices and also plant protection measures ensure bumper yields.

only from the fertility of the soil, but from ancillary factors like availability of moisture for producing the crops, facilities for drainage, lack of harmful conditions in the soil, etc. A fertile soil may not be a productive one owing to certain limitations such as, lack of irrigation facilities, the presence of toxic constituents or limitations of soil depth, etc. A fertile soil may contain all the essential nutrients needed for plant growth, but due to unfavourable soil conditions of, say, the pH conditions, the plants may not be in a position to utilise them for good growth. Thus, soil productivity is a function of environmental factors combined with soil fertility, or, more correctly, we can state that soil fertility in combination with environmental factors and management practices constitutes soil productivity.

Soils which are inherently of low fertility can be made productive through adequate management measures. Thus, the application of the necessary fertilizers and manures can make even soils

FIG. 3. Land with low natural fertility can be made highly productive through fertilization. High yields can be obtained with fertilization (crop in rear) while the farmer's practice (farmyard manure only—foreground) can only give a poor yields from the same land.

with low fertility highly productive. Such improvements in the productivity of soils through human effort and management can frequently be observed in the vicinity of large urban areas where poor land is usually developed by human effort to produce a wide variety of vegetable and horticultural crops on a highly paying basis. Soils which are inherently of high fertility but not adequately productive because of adverse environmental conditions can also

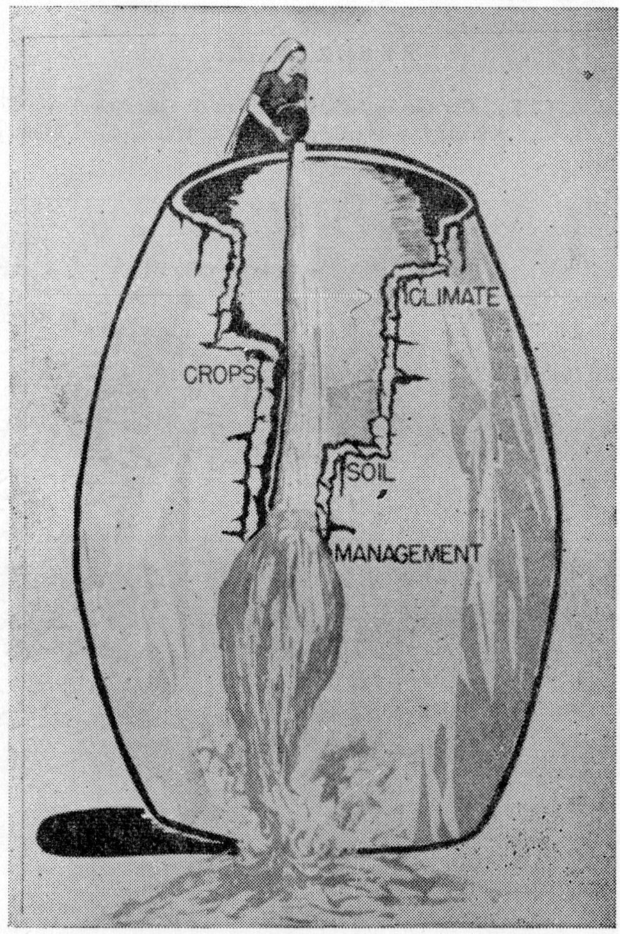

FIG. 4. Management of the land is a very important factor controlling productivity. Inputs through various measures to increase production can be lost if the management is low.

be made productive through proper management measures. An important means of doing this is through provision of adequate water supplies needed by the crops and through irrigation facilities. Other factors which restrict crop production, like drainage impedance, restriction of plant root penetration or toxic soil conditions, can be overcome through adoption of suitable management practices or amelioration measures that overcome these handicaps.

REFERENCES

Kellogg, C. E., *Our Garden Soils*, Macmillan & Co., New York, 1952.

Jacks, G. V., *Soil and How to make the most of it*, Philosophical Library, New York, 1954.

CHAPTER II

THE GOOD SOIL AND THE BAD SOIL

DEPENDING upon their qualities and properties for raising crops satisfactorily, soils may be divided into two groups to describe them in terms of popular notions, as good and bad (or poor) soils. Such a division may be affected by visual examination, but it takes long experience and skill to judge and describe a soil as being either good or bad. Even such a judgement can only be of a general nature and may not be indicative of how the soil will react, for example, to irrigation, to some specialised cropping or intensive cultivation. A faulty judgement, not backed by sufficient scientific information, can lead a farmer into difficulties and uncertainties which are avoidable. Modern methods based on scientific assessment of the various factors of the soil are available to assess the characteristics and potentialities of any soil, and these deserve to be availed of to prevent gross mistakes which can result from erroneous judgement.

FIG. 5. Soil is formed under a complex of dynamic factors.

It is, therefore, of great advantage to obtain advance knowledge about the important characteristics that distinguish the good soil and help in its differentiation from the bad or poor soil. It is, however, pertinent to indicate that natural soil is a complex

of various dynamic factors which are continually changing and are bound to change under field conditions. As such, a bad soil need not always remain a bad soil, simply because it happens to have, at the time or period of observation, certain unfavourable factors affecting it. There are many ways of altering or bringing about a change in t e unfavourable factors affecting a soil, rendering it a bad or poor soil, and converting it into a good or productive soil. We may enumerate here some of the important factors that determine or control the desirable qualities of the soil of agricultural lands. Of these, topography of the terrain, texture, structure and depth of soil and the associated factor of drainage are important factors of a physical nature controlling the desirable properties, while soil reaction, presence or absence of harmful salts, supply of plant nutrients, and availability of soil organic matter are the major chemical and physico-chemical factors controlling these properties. The role of each of these factors in influencing the qualities of the soil can now be briefly discussed.

TOPOGRAPHY

Of these various factors, topography is an important one which decides the behaviour of a soil under cultivation and its reaction to agricultural operations. A soil, which occurs on level land, will usually be more favourable for agricultural operations and use, compared to another soil located on steep sloping ground where there is a greater scope for accelerated erosion and loss of fertile soil from the land surface. Soil in level locations has a greater chance of being homogeneous, more uniform in physical characteristics, and relatively more easy to manage under cultivated conditions. Moisture received by the soil through rainfall or irrigation and the fertilizers applied on them would tend to be retained in the soil a longer time without being rapidly lost either by leaching, or surface run-off. Cultivation and other agricultural operations can also be easier and kept at a minimum. On the other hand, lands on steep slopes tend to have heterogeneous soils owing to lateral movement of the surface particles, likely exposure of the subsurface layers, and also gullying by removal of soil layers through erosion. Such soils invariably require intensive soil conservation measures to keep them in proper condition. Adequate and normal drainage is a prerequisite for successful cropping. While in a level or moderately sloping topography, provided texture and depth of

FIG. 6. Soil on sloping land is liable to be washed away easily. Soils in such locations are difficult to cultivate and require intensive management.

soil are satisfactory, drainage can be expected to be adequate for normal cropping or can be controlled without great difficulty in land with steep slopes, it can easily become excessive and give rise to a variety of problems. Moisture retention, its availability to growing crops, and droughtiness, all of which are influenced by topography, can become limiting factors affecting the crop production. Also intensive management practices will become necessary, depending upon the topographical factors, to conserve the moisture, the soil itself, and its fertility.

TEXTURE

Soil texture is the size distribution of primary particles which compose the soil. Primary particles are the individual mineral grains of the soil. The soil particles are usually distinguished as coarse and fine sand, silt, and clay. The texture is due to different proportions of these particles in the soil. The texture of a soil is very important in determining its physical behaviour. Many chemical and biological characteristics of soils are also related to soil texture. The texture of the surface soil determines to a large extent the tillage properties of the soils. A clayey soil will render the field too sticky when wet and too hard when dry, and the field operations, therefore, under both the conditions will be more difficult compared to a lighter textured soil. Clayey soils, because of their moisture retentivity, will, however, be suitable for some type of crops which require a great deal of water for their growth. In

FIG. 7. Puddling soil to prepare the land for paddy cultivation. Soil structure is destroyed in this process.

particular, for example, for a crop like paddy, which requires a puddled soil and one which can retain a great deal of water, a clay soil possesses advantages over a light textured soil that does not have these desirable properties. If the soil is sandy or light textured, it may be easy to work over a wide range of soil moisture conditions, but the moisture retentivity will be poor and so droughty. The kind of crops that are grown on such soils may be few and the variety limited. Under irrigated conditions in such open textured soils, added fertilizers may also be washed down rather easily and lost to the crops. This textural property, however, has advantages for certain types of crops like potatoes, carrots, and groundnuts, where open texture is desirable because of the facility afforded by such soils for their good growth. Again, proper drainage, which is normally demanded in all cultivated soils, is controlled by texture of the surface and of the sub-surface layers. The drainage is limited or moderate in clayey soils and this property, as indicated above, may be beneficial to certain crops but not to others. Similarly, in the case of light soils, with an open texture, drainage is likely to be free or excessive leading to problems of moisture limitations and nutrient supply. A moderately well-drained condition, provided other conditions are satisfactory, is associated with a soil of a loamy texture, and this condition can facilitate the necessary movement of air in the soil, and ensure subsoil aeration, which is a factor as important for satisfactory crop growth as moisture availability. A soil which is neither too clayey nor sandy will normally be preferable for cropping, and will be suitable for a wide variety of crops. By experience, it is observed that a soil with a loamy or sandy loam texture and possessing a clay content between 10 and 30 per cent, and a similar range of silt content, is the most desirable for most kinds of general crops.

STRUCTURE

Soil structure refers to the aggregation of primary soil particles into compound particles. Soils made up of practically all sand or silt do not show any structural arrangement because of lack of binding properties of clay. A well-developed structure usually indicates the presence of optimum content of clay. A good crumb structure which can help to retain sufficient moisture, and yet allow sufficient aeration and root growth, is the most desirable property in soils. However, this may not always be necessarily so. For example, paddy may need a puddled soil where the structure is lost in the process of tillage operations, and yet the soil functions well for this crop. In the poorer sandy soils, the soil particles may be so large, and the binding substances and the finer constituents so low, that soils will not form definite aggregates which are necessary for a

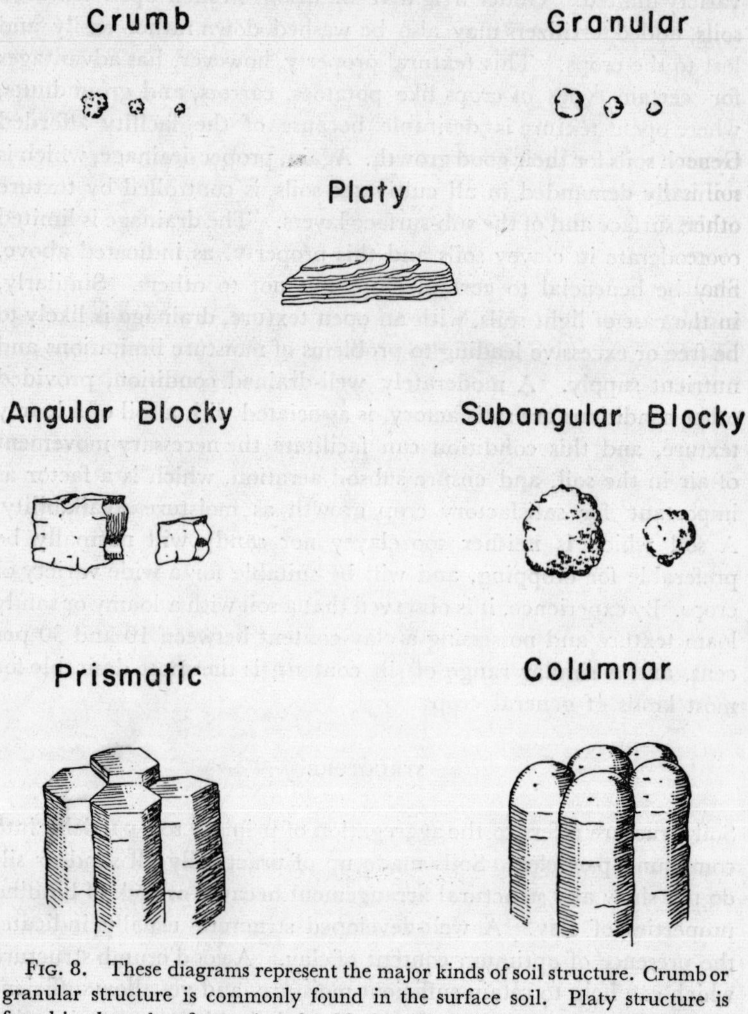

Crumb
Granular
Platy
Angular Blocky
Subangular Blocky
Prismatic
Columnar

FIG. 8. These diagrams represent the major kinds of soil structure. Crumb or granular structure is commonly found in the surface soil. Platy structure is found in the subsurface. Angular blocky, prismatic, and columnar structures are found in the subsoil.

good structure. In the absence of such a satisfactory structure, the soils suffer from the disadvantage of droughtiness and the incapacity to bear heavy yields of most varieties of crops. The structure of soils is usually the resultant product of soil management and cultural operations in cultivated lands, and structure improves in a wide range of soils with tillage, manuring, and cultivation practices.

SOIL DEPTH

Generally, in the field under natural conditions, the depth of the soil is variable. It may be only a few inches in certain fields and in others go down several feet. A shallow soil can bear only shallow rooted crops and such soils are not suited to deep rooted crops. Shallow soils have only limited sources of nutrient and moisture supply, and they tend to erode easily by wind and water action. Management of such soils requires great attention and effort. A deep soil, on the other hand, is usually more homogeneous, tends to be more stable, has large sources and reserve for supply of nutrients, and can retain moisture for longer periods. The deep soils are suited for a wide variety of crops both of the annual and perennial varieties. Included in the latter category will be fruit trees and orchards, which normally cannot thrive under shallow soil conditions. The shallow types of soil are, generally, found on steeply sloping lands, along hillsides, open wind-swept plains or other areas susceptible to high erosion, and on rocks difficult to weather. Deep soils are commonly found in valley areas, wide plains, and the deltaic regions.

ORGANIC MATTER CONTENT

The organic matter content of soils plays a very important part in determining the qualities of a soil. Rich soils are distinguished by their high percentage of organic matter content. Organic matter helps to improve the physical properties of a soil like structure, water-holding capacity and tilth. The soil is made spongy, aeration of the soil and subsoil is facilitated, as also drainage. It also promotes greater biological activity in the soil leading to better nitrification, nitrogen fixation, and other favourable processes. Organic matter also tends to control and limit development of harmful alkaline conditions by neutralising the effects of some of the alkaline salts. Freshly decomposing organic matter is also a

Fig. 9. A deep soil on a level or gently sloping terrain is easy to cultivate, has large sources of supply of plant nutrients and also moisture. A wide variety of crops can be grown on them.

source of nitrogen supply for plants. Soils containing less than one per cent of organic matter are generally considered to be poor in fertility and may need to be supplied with extra applications of organic manures to bring up the organic matter content to satisfactory levels. Under tropical conditions organic matter is lost from soil on account of the high atmospheric and ground temperatures combined with the high biological activity. Various workers have studied the influence of climate and vegetation on the organic matter content in Indian soils. Hans Jenny, *et al.*, in a detailed study of soils from over 500 locations distributed in the different climatic regions of India found that primarily environment and, secondarily, cultivation cause depletion of organic matter in soils which rapidly stabilize on a low content level, i.e. between 30 and 60 per cent of the original.

The organic matter lost in soils is replaced through the application of well-decomposed manures like farmyard manure, and urban or rural composts. For most dry land crops the application of such manures, annually or at least periodically, is necessary for ensuring normal crop yields. However, paddy thrives well when the soil contains large amounts of freshly decomposing organic matter. High organic matter in soils is also found necessary for crops like betel vine, mushrooms and certain vegetable crops and plantation crops like areca, coffee, and tea.

BIOLOGICAL CONDITION

Good productive soils possess high micro-biological activity. Such activity promotes the development of proper conditions necessary for root growth and effecting the chemical changes in the soil which can ensure the optimum conditions of temperature, release of mineral nutrients to the plants, and the absorption of these nutrients by the plants. Microbiological activity ensures the release of nutrients from soil minerals, from the organic matter present in the soil or from added fertilizers, besides helping to fix nitrogen of the atmosphere into compounds which can be useful to the plant. In fact a measure of the biological activity in the soil is sometimes taken to indicate its fertility level. Virgin soil, desert soils, soils in fields, which have been newly terraced, burnt soil, and acid, saline and alkali soils are always low in biological activity and, hence, considered relatively infertile. When such soils are brought under cultivation for the first time or after a long fallow period, they behave as relatively infertile soils, and even with fertilizer applications react only slowly on account of the low biological activity. It is because these soils lack in the micro-organisms which can work on and transform the constituents of the fertilizers into forms absorbable by the plant roots, it will be necessary to allow a short period to lapse under cultivation for the soil to be rehabilitated and to enable the establishment of proper biological conditions. This process of rehabilitation can be speeded up through innoculation of the soils artificially with the requisite types of bacteria through special cultures. Normalising the soil conditions may require reclamation operations in some cases and, as demanded by the situation, resort will have to be made to liming and phosphating the soils and also raising leguminous crops.

The normal biological activity in cultivated soils may be hindered under certain other conditions by the toxic or unfavourable conditions developing in the soil as a result of impeded drainage, or continued cultural practices of an unsatisfactory nature involving use of manures of an unsuitable or unbalanced nature. The use of toxic chemical insecticides and fungicides, after prolonged periods, can also result in reduction of the biological activity in soils where they were previously normal. In all such cases, rehabilitating the soils to normalise the biological conditions will be a prerequisite for satisfactory and successful cropping.

DRAINAGE

Free movement of water through the soil in cultivated fields is essential for maintaining the soil in a fertile condition. Such movement provides for removal of toxic substances from the root zone and in helping the circulation of fresh soil solution and soil air through the subsoil layers. The absence of such free movements, or their reduction through impedance of the drainage conditions, brings about various harmful conditions in the soil which ultimately affect plant growth.

Stagnation of water received through rainfall or irrigation leads to the development in the soil of an environment unfavourable for normal root life and plant growth. Toxic conditions rapidly develop which result in the release of harmful compounds and gases in the subsoil layers which, while proving immediately toxic to the plant roots, steadily injure the crop and eventually render the soil conditions unfavourable to plant growth. Free movement of air and oxygen through the soil is essential for the healthy growth of plants and this is arrested when drainage is impeded. Saline and alkaline conditions may also develop under certain situations of impeded drainage. Also, the soil structure may be harmed. Under conditions of drainage impedance and moisture stagnation, the microbiological activity in the soil is lowered. All

FIG. 10. A well-drained land is suitable for cultivation of a variety of crops. Good drainage is dependent upon the topography besides the texture and profile characters of the soil.

these harmful conditions in a cultivated field react in giving poor crop yields, or their total failure and, when the conditions are extreme, possibly in ruining of the land. In cultivated soils, therefore, there is great need for maintenance of satisfactory drainage conditions to ensure the optimum physical and physico-chemical conditions, which in their turn control the proper health of the soils so important for maintaining them in productive condition.

Drainage is dependent, as described earlier, upon the texture of the soil, its structure, permeability of the subsoil, and topography of the land surface. Drainage is always poor in clayey soils, in soils with high sodium in the exchange complex, and in soils having a hard or impermeable pan or rock formation below the surface soil. Defective drainage and waterlogging of soils are common in low-lying lands and in lands lying in hollows (where the scope for outlet of surplus water is poor), and in fields lying below tanks and irrigation canals. While impeded drainage is harmful, excessive drainage, on the other hand, is also a disadvantage. Excessive drainage leads to the washing away of the finer particles of soil and the nutrients from the soils, without which, the soil will tend to be droughty. In consequence, highly drained soils, such as sandy soils or laterite soils, are poor in fertility and demand a great deal of attention to maintain the moisture regime. A soil with a good or moderate drainage is, therefore, to be desired from the point of view of its productivity. Extensive areas in different states in India are found to be affected by a high water table, resulting from waterlogging. Table 1 gives the approximate estimate of these areas. It is seen that over 8.5 million acres are affected by this disadvantageous condition.

SALINE AND ALKALI SOIL

In India, saline conditions have developed through various means, some natural and some man-made. Large areas of saline and alkali soils are found to occur naturally in Rajasthan and in many areas in the Punjab, Uttar Pradesh, and in the central areas of the Deccan plateau after the introduction of irrigation. It is estimated that over 20 million acres of such land exists in our country and this acreage is increasing year by year. Because of unfavourable soil and climatic conditions, harmful concentrations of certain salts accumulate in some places either in the top soil or in the subsoil. In such soils, crop growth is affected through failure of a proper germination of the sown seed, and even if the germination is satisfactory it subsequently leads to poor growth of the crop or results in total failure. The salts, which usually accumulate in large concentrations in soils, are the carbonates, chlorides and sulphates of sodium, calcium and magnesium, and, in any area affected by high salinity,

TABLE I*

EXTENT OF WATER-LOGGED AND SALT-AFFECTED AREAS IN INDIA

(*in '000 acres*)

State	*Category A 0-5' deep*		*Category B 5-10' deep*	*Category C 10-15' deep*		*Category D Salt-affected area in deep water table*	*Total waterlogged area*	*Total salt-affected area*
	Water-logged	*Salt affected*	*Water-logged*	*Water-logged*	*Self-effected*			
Punjab	1,971	300	3,350	450	2,283	1,250	7,604	3,000
W. Bengal	574		191				765	
U.P.		n.a.	96	n.a.		2,300	96	2,300
Bombay	11	18		54	n.a.		11	72
J & K	5		25		24		54	
New Delhi	3		6		1		10	
	2,564	1,318	3,668	504	2,308	3,550	8,540	5,372

n.a.=Not available.

*P.C. Bansil, *Journal of Soil and Water Conservation in India*, Vol. 8, No. 4 (1960), 14.

these salts individually or in combination may be present. A saline soil has excessive amounts of soluble salts only. An alkali soil, on the other hand, has excessive amounts of adsorbed sodium which may form more than 15 per cent the exchange complex. Soils which are saline or alkaline tend to be difficult to cultivate, and responses of crop growth to fertilizer applications are poor unless they are improved, and the harmful conditions remedied through reclama-

FIG. 11. Salinity-affected land. The surface soil is covered with salt incrustation. Excessive salts render soils infertile and difficult to cultivate.

tion. The methods of reclamation involve the removal of the harmful concentrations of salt through leaching with large quantities of water, providing for adequate drainage and adoption of ameliorative measures including the application of sulphur, gypsum, and organic manures. Saline and alkali soils are generally found in conditions of impeded drainage, in areas where the soils are inherently rich in these salts, or in arid regions where atmospheric precipitation is not enough to remove them from the soil layers. In other soils, the salts which may be present in high concentrations at a depth below the layers, where the plant roots can be affected, are brought up by capillary action or by the upward movement of the soil water after the introduction of irrigation by artificial means. The influence of saline and alkali conditions on the growth of paddy and other crops under Indian conditions has been extensively studied.

The general findings of Agarwal and Yadhav[1] are that paddy is highly tolerant towards alkalinity as indicated by high a pH and also high salt content. Other crops are not so tolerant towards high pH and their growth is also very severely affected with increasing salt concentration in the soil. Paddy can grow normally in soils with pH up to 9.0, the salt concentration, however, remaining under 150 parts per million. With higher concentration of salt (up to 3,000 ppm) at this pH limit, the growth is considerably stunted, and above 3,000 ppm the crop does not grow. With other crops the pH level for satisfactory growth is lower than the above-mentioned pH limit and, over pH 8.2, there is no satisfactory growth. Even at this level of pH and a salt concentration of between 150 and 3,000 ppm, the growth is considerably stunted, and with higher salt concentrations the growth is altogether suppressed.

SOIL REACTION

Soil reaction is a measure of the acidity or alkalinity of the soil. Acidity or alkalinity is measured by the range of pH which extends from 0 to 14. A pH of 7 is said to be neutral and a figure of pH from 0 to 7 indicates acidity and from 7 to 14 alkalinity. Crops grow best only under a limited range of soil reactions, viz. between pH 5.5 and 8.5. A few crops may grow well in acid soils with pH between 5.0 and 6.0, while some may even tolerate a pH lower than 5.0. Similarly, a few crops grow best in alkaline conditions, say between 7.5 and 8.5, while some crops stand alkalinity higher than 8.5. Beyond these ranges of acidity and alkalinity, however, soil productivity weakens and crop growth is severely affected in extreme cases. Table II gives the range of pH, most favourable for the growth of different common crops.

[1]R.R. Agarwal and J.S.P. Yadhav, *Journal Indian Society of Soil Science*, Vol. 4, No. 3, 1956.

TABLE II

MOST FAVOURABLE SOIL REACTION FOR GROWTH OF DIFFERENT CROPS*

Crop	*Range of pH*	*Crop*	*Range of pH*
Barley Hodeum vulgare	6.5—8.0	Field beans Phaseolus vulgaris	6.0—7.5
Maize Zea mays	5.5—7.5	Clover (berseem) Trifolium alexandrinum	6.0—7.5
Rice Oryza sativa	5.0—6.5	Cowpea Vigna sinensis	5.0—6.5
Wheat Triticum sativum	6.0—7.5	Groundnut Arachis hypogea	5.3—6.6
Upland Cotton Gossypium hirsutum	5.0—6.0	Lucerne Medicago sativa	6.2—7.8
Flax Linum usitatissimum	5.0—7.0	Peas Pisum sativum	6.0—7.5
Hemp Cannabis sativa	6.0—7.0	Soya Beans Glycine maxima	6.0—7.0
Potatoes Solanum tuberosum	4.8—6.5	Hariali Grass Cyndon dactylon	6.0—7.0
Sweet Potatoes Ipomaea batatas	5.8—6.0	Johnson Grass Sorghum helepense	5.0—6.0
Rape Brassica rapus	6.0—7.5	Jowar Sorghum vulgare	6.0—7.5
Sugarcane Saccharum officinarum	6.0—8.0	Italian millet Setaria italica	6.0—6.5
Tobacco Nicotiana tobacum	5.5—7.5	Sudan grass Sorghum vulgare var. Sudanese	5.0—6.5
Plantains Musa spp.	6.0—7.5	Grapes Vitis vera	6.0—8.0
Tea Camellia sinensis	4.0—5.5	Tomatoes Solanum lycopersicum	5.5—7.0
Rubber Hevea braziliensis	3.5—8.0		

*Extract from *Efficient Use of Fertilizers* (Ed. Ignatief), F.A.O., 1958.

TABLE II (*Continued*)

Crop	*Range*	*Crop*	*Range of pH*
Coconut		Cabbage	
Cocos nucifera	6.0—7.5	Brassica var.	5.5—7.5
Coffee		Onions	
Coffea arabica		Allivium cepar	6.0—7.0
Coffea robusta	4.5—7.0		

Acid soils in India are usually found in the humid forest regions where excessive rainfall and vegetation growth have tended to the removal and depletion of calcium from the soils, leading to acid conditions. Such soils exist in the Western Ghats, along the West Coast, West Bengal, Assam, and the Himalayan region. Extremely

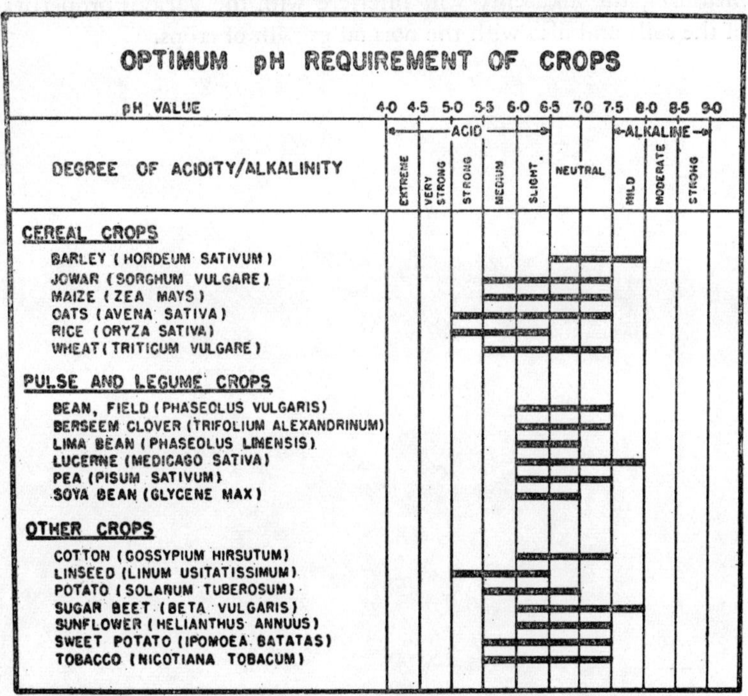

FIG. 12. The optimum ranges of soil pH for the growth of different crops in India.

acidic soils, with pH as low as 4.0, called the "Kari" soils are found in the coastal areas of Kerala. Acid soils, besides being low in base content, and particularly calcium content, contain toxic concentrations of aluminium and iron. These soils suffer from poor biological conditions, low phosphate status, and respond poorly to cultivation and to fertilizer applications. Unless they are limed to neutralise their acid condition and otherwise treated to improve their nutrient status and the biological condition, they are relatively infertile soils.

The harmful conditions of high alkalinity in soils have been dealt with in the earlier paragraphs. Soils of high alkalinity having pH of above 8.5, usually, contain excessive amounts of sodium which tend to harm soil structure, cause deflocculation of the clay, and interfere with root growth, besides affecting the availability of plant nutrients like nitrogen and phosphorus to the plants. Unless these harmful conditions are remedied through suitable reclamation measures, the alkalinity will interfere with the various properties of the soils and also with the normal growth of crops.

CHAPTER III

THE PHYSICAL MAKE-UP OF SOIL

PHYSICAL CONDITIONS OF THE SOIL

The physical properties of a soil are very important in influencing its productivity, and effective management is needed to maintain it at a high level. Like its other properties, physical conditions of a soil are dependent upon several internal and external factors and they are constantly changing under natural field conditions. The favourable physical conditions of a good, productive soil are generally built up with great care over an extended period of time and require constant attention for their maintenance.

The basis of the physical properties in soil is the texture, which in turn is dependent upon the amount and the kind of individual mineral particles. The mineral particles in the soil are of all sizes and shapes, and range from particles visible to the

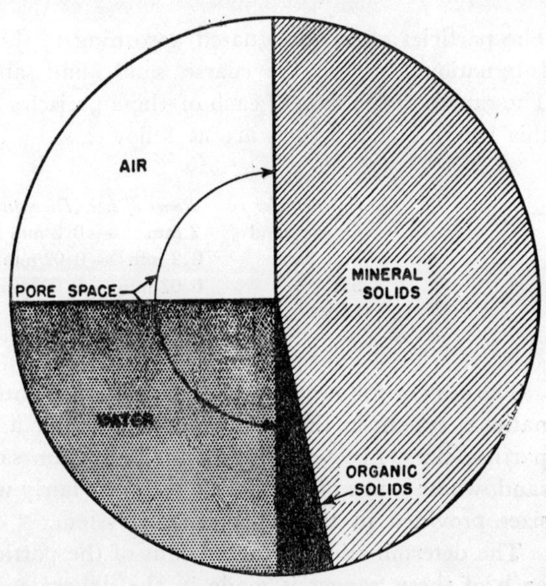

Fig. 13. The volume composition of a loam soil in good tilth is approximately 50% solids (mineral + organic) and 50% pore space. When the soil moisture is adequate for normal plant growth, approximately one half of the pore space will be filled with water and the other half will remain as air.

naked eye down to those which can be seen only with the aid of a high-powered electron microscope. For practical use, however, these particles are graded into ranges of different sizes, and the proportional combination of these different-sized particles gives the textural classification of the soils. The three main groups of soil particles popularly known are sand, silt and clay, in the descending order of the sizes of the particles. In nature, however, they are found in a large number of combinations that yield a wide range of textures like sandy loam, clay loam, sandy clay, silty clay, etc., falling between the extremes of sand and clay. A loam is a popular term used to indicate a soil having intermediate properties between a clay and sand. It has roughly equal proportions of the sand, silt and clay fractions.

SOIL PARTICLES — MECHANICAL ANALYSIS OF SOILS

The particles of soil designated according to the standards of the International system are coarse sand, fine sand, silt and clay. The ranges of the sizes of each of these particles as accepted under this International system are as follows:

Soil Particles	*Range of size (Diameter)*
Coarse Sand	2 mm — 0.2 mm
Fine Sand	0.2 mm — 0.02 mm
Silt	0.02 mm — 0.002 mm
Clay	Less than 0.002 mm

The system followed in America differs slightly from the International system in that provision is made for a larger number of particle sizes, like very coarse sand, medium sand and very fine sand, whose range of sizes falls within the fairly wide range of sand sizes provided in the International system.

The determination of the amount of the particles falling within each of these ranges is made in the laboratory by the methods described as Mechanical Analysis of Soils. A number of methods is available; some of them are detailed and time-consuming, while others, though quick, give results within a reasonable degree of accuracy. Based upon the content of the different proportions of these particles, the textural class of soil is derived with the help of the textural diagram. This diagram which provides for variations in the contents of sand (consisting of coarse plus fine sand percentage), silt and clay is reproduced in Fig. 14.

TEXTURAL CLASSIFICATION IN THE FIELD

For practical purposes it is quite adequate to adopt simple methods to judge the texture of the soils in the field itself. With

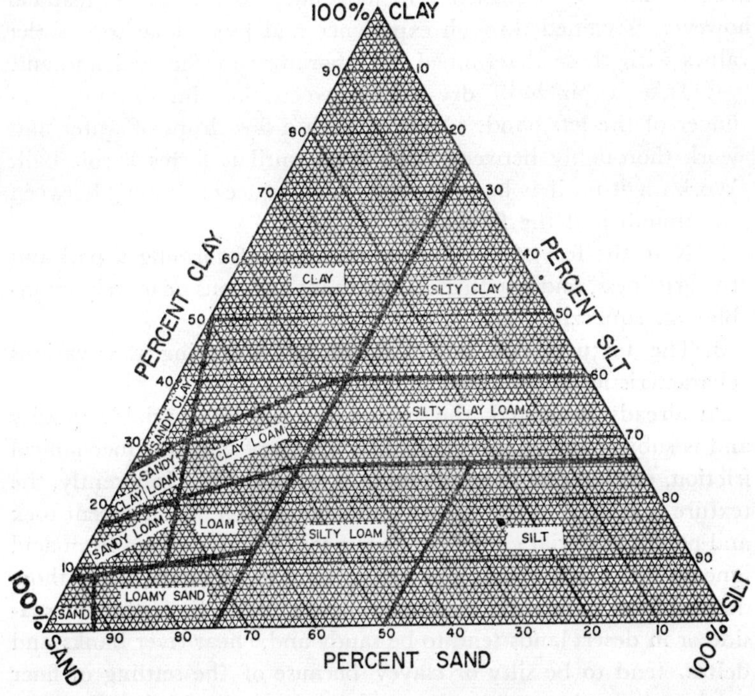

FIG. 14. Triangular texture diagram based on international fractions, with effective diameters of 0.002, 0.02, and 2 mm for the upper limits of the clay, silt and sand fractions respectively.

a little practice it is quite possible to do this with a fair degree of accuracy with any soil by a feel of a small mass of the moistened soil rubbed between the thumb and the forefinger. While the laboratory analysis gives accurate figures for each one of the component soil particles, from which the texture can be interpreted, a ready determination of the texture in the field itself helps to decide on this property on the spot. This field method is described here in brief.

DETERMINATION OF SOIL TEXTURE

Feel Method

Soil texture is determined by the feel of the soil when it is rubbed between fingers. Proficiency in judgement of the correct texture, however, is gained through experience and by comparison of the values with those determined by laboratory mechanical analysis.

1. Take a pinch of dry soil between the thumb and forefinger of the left hand. Moisten with a few drops of water and work thoroughly between the fingers until it forms a soft ball. Work on it until it becomes stiff; then squeeze it out between the thumb and the forefinger.
2. Note the feel of the fingers, the ease of forming a ball and the grittiness, and whether soil turns into ribbons or merely crumbles on squeezing, etc.
3. The textures can be classified by observing the various characteristics as indicated in Table III.

As already indicated, the texture of a soil is a variable quality and is subject to changes caused by effects of climate, mechanical friction, erosion due to wind and rain, tillage, etc. Inherently, the texture is related to the nature and composition of the parent rock and parent material giving rise to the soil. Soils derived from acid igneous rocks rich in siliceous matter tend to be sandy, while those derived from basic rocks tend to be fine-textured. Soils near the seaside or in desert lands tend to be sandy and, near river banks and deltas, tend to be silty or clayey because of the settling of finer particles from the muddy waters of the rivers. It is, therefore, not practicable to attempt to change the texture of a soil to suit one's requirements except on a very limited scale. It would be advisable to have a cropping pattern suited to the prevailing conditions of the texture of the soil rather than try to build up a different texture. A sandy loam may be cultivated for crops like groundnuts, potatoes, or millets which do well in open textured soils, whereas a heavier soil may be taken up for crops like paddy, wheat, etc.

TABLE III

DETERMINATION OF SOIL TEXTURE (FEEL METHOD)

Textural class	*Feel of fingers*	*Ball formation*	*Stickiness*	*Ribbon formation*
Sand	Very gritty	Does not form a ball	Does not stain finger	No
Loamy sand	—do—	Forms a ball which is easily broken	Stains finger slightly	—do—
Sandy loam	Moderately gritty	Forms a fairly firm ball but is easily broken	Stains fingers	—do—
Loam	Neither very gritty nor very smooth	Forms a firm ball	—do—	—do—
Silt loam	Smooth or slick "buttery" feel	—do—	—do—	Slight tendency to ribbon with flaky surface
Clay loam	Slightly gritty feel	Moderately hard ball when dry	—do—	Ribbons out on squeezing but ribbon breaks easily
Silt clay loam	Very smooth	—do—	—do—	Shows some flaking on ribbon surface similar to silt loam
Clay	—do—	Forms a hard ball which when dry cannot be crushed between fingers	—do—	Squeezes out at right moisture into long (1″-3″) ribbons

STRUCTURE

The structure of a soil defines the aggregation of soil particles. The soil by its nature contains a number of components, viz. mineral particles, soil moisture with soluble matter dissolved in it, air, micro-organisms, organic debris like decomposing roots, animal remains, etc. These individual components may exist under a variety of combinations and exert their own individual influences upon the physical properties of the soil. The structure of the soil is determined by the combined force of these effects and is further subject to the modifying influences of external forces like rainfall causing leaching, erosion, waterlogging and cultural practices including tillage and manuring.

A favourable soil structure is necessary for proper root growth of most types of crops and for providing the necessary aeration and drainage to keep the soil in a condition fit for plant life. In order to distinguish between good and bad soil structure, certain classifications have been made and their chief properties have been defined so that by an examination of soil structure, one can obtain a rough idea of the nature of the soil, its suitability for certain crops and its

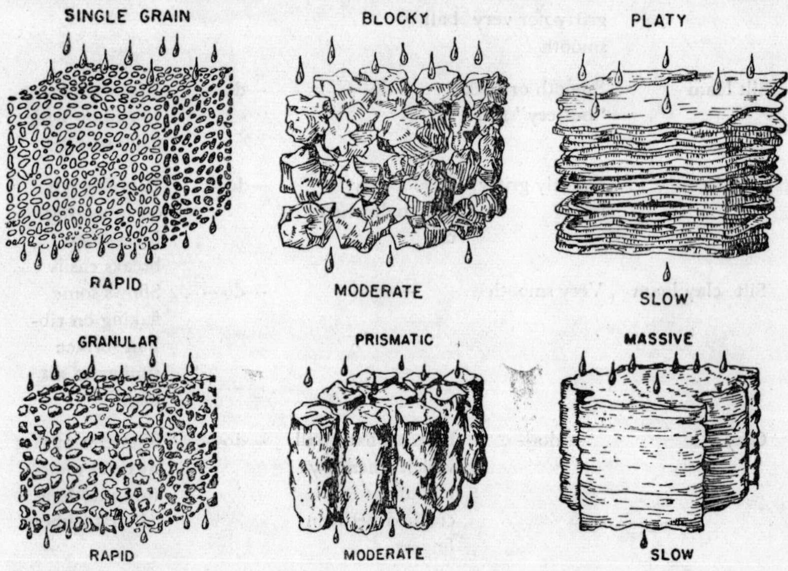

Fig. 15. The generalised relationship between soil structure and the infiltration rate of water into the soil. A soil with a single grain or granular structure, has a rapid infiltration rate; a soil with a platy or massive structure has a slow rate of infiltration. In rice paddy soils there is a conscious effort to create a platy or massive structure at a depth of approximately 6 inches to reduce the loss of water by percolation. For all up land crops, it is desirable to so manage the soil that it has a moderate to rapid infiltration rate to reduce run-off losses of water and to reduce soil erosion.

behaviour under cultivation and cropping. The major kinds of structure commonly met with in soils are depicted in the illustration (Fig. 15). The following chief classes of structure are commonly recognised and the properties associated with each are given below in Table IV.

TABLE IV

Structure	*Description*	*Property*
Platy	Natural cracking is horizontal. The horizontal plates of soil overlap, making it difficult for air, water and plant roots to penetrate the soil. Found in the soil subsurface	Has poor permeability and is swampy
Prismatic	Cleavage is vertical. Found in B horizon of loamy soils	Easy permeability and good drainage
Blocky	Both angular and subangular blocky aggregates whose corners are sharp angular or rounded. Found in the subsoil	Excessive drainage
Columnar	These aggregates are longer than wide and usually have rounded tops. Found in the subsoil	Fairly easy permeability and moderately good drainage
Granular	They are similar to crumb, but the aggregates are larger and of different sizes. Do not have sharp edges; smooth surfaces. The aggregates appear harder and less porous than crumbs. These also are found in the surface soil	Ideal for root growth
Crumb	The aggregates are small and held together weakly. They are porous, like crumbs of bread and possess plenty of air spaces. They are commonly found in the surface soil	Ideal for root growth
Massive	Compact and uniformly packed. Little air space, and movement of air and water difficult. Found commonly in the subsoil	Poor drainage and low permeability

For the best growth of crops, soil with a structure, which is favourable for adequate drainage, free movement of air and penetration of roots is essential. Such structure is normally obtained with a soil having a crumb structure in the top soil with a blocky, prismatic or columnar structure in the subsoil. The structure of the surface soil, if it is crumb, will suit most types of crops. This structure is promoted by good tillage practices. A crumb is formed by the binding effect of colloidal organic matter and various other known and unknown factors, like presence of lime, colloidal clay, etc. One of the ways of promoting crumb formation in soils is to add organic matter by the application of green manures, cattle and farm manures. Fallowing and at times, turning the land under grass helps crumb formation in soils. Repeated or continuous cropping without taking adequate measures to rest the soil or without adding organic matter to permit humus formation which helps proper aggregation of soil particles, causes loss of the desirable structure.

Soil structure is also destroyed by the loss of organic matter, or by accumulation of sodium in the exchange complex, or by flooding and puddling, and by too much of tillage. Good structure is not required equally by all crops. However, each crop grows best only when the soil has certain structural characteristics best suited for its requirements. Some crops like paddy grow in puddled soils where structure is lost or destroyed through puddling operations.

REFERENCES

ROBINSON, G.W., *Soils, their Origin, Constitution and Classification*, Thomas Murby & Co., London, 1936.

HILGARD, E.W., *Soils*, Macmillan & Co., New York, 1906.

RUSSELL, E.W., "Soil Structure," *Imp. Bur. Soil Sci. Tech. Com.*, No. 37, C.A.B., London, 1938.

BAVER, L.D., *Soil Physics*, John Wiley & Sons, N.Y., 1956.

CHAPTER IV

SOILS AND THEIR DEVELOPMENT

Soil can be defined as the thin mantle of weathered rock material covering the surface of the earth's crust and which can sustain plant growth. The depth of soil on the land surface varies from place to place and ranges from a few inches to several hundred feet in thickness. It is found to occur in thin layers over rocky ground and in great thickness as alluvial deposits along river courses and deltas and in wind-blown deposits as continental loess. However, when compared with the gigantic size of the earth, which is almost 8,000 miles in diameter and 25,000 miles in circumference, the soil layer can at best be described as a very thin skin of the earth's crust. To get a rough idea of the relative sizes of these, we may compare the earth to a gigantic potato and the soil as its thin outer skin. As regards the size of the potato in this case, this would be several million times bigger than the natural tuber, with the thickness of the skin remaining the same. The above comparison is made to give an idea of the relatively thin surface of the crust which has decomposed and become the soil, which sustains the world's human and animal population, compared to the vast mass of the earth which has not been affected by these processes.

The thin covering of soil on the earth's crust is formed from the underlying bed rocks by the process known as "weathering." This is a process which at best is a combination of physical, chemical and microbiological reactions causing the decomposition of rocks and of plant and animal materials. If the processes of operation are largely physical, as may be true in dry and arid regions, the soil will resemble the parent rocks very closely in its chemical and mineralogical composition, the differences being largely a matter of fineness of division. If in addition, the rock particles are subjected to the solvent and leaching action of water containing carbonic and other acids and various salts in solution, as occurs in humid regions, deep-seated changes will be effected. Under these conditions, the inorganic part of the soil may become largely an accumulation of finely divided, relatively insoluble mineral residues that bear little resemblance to the original rocks. This weathering process is a gradual one and the formation of soils from the parent rocks is very slow. It is estimated that it may take anything from a few hundred to a thousand years to form an inch of top soil depending upon the intensity of the weathering action.

The original source of the inorganic constituents of soils are the rock materials of the earth's surface. The kind of rock materials present at any location on the earth's surface is determined by the types of geological activities that have taken place. Rock materials vary widely in their composition and make-up, and accordingly they are named differently to characterise them. All rocks are made up of silica (quartz) and various minerals in which different elements occur in chemical combination with one another, and also with silica as silicates. The important minerals that occur in different kinds of rocks and some of the characteristics by which they may be identified are listed in Table V.

TABLE V

SOIL MINERALS

Sl. No.	*Name of mineral*	*General appearance*	*A principal constituent of*
1	Quartz	colourless, white, smoky or pink; glassy	granite, sands and silts, quartzite, sandstone
2	Feldspars	white, gray, pink or reddish	granites, gabbros, felsites, basalts
3	Augite and Hornblende	dark-green to black prismatic crystals	gabbros, basalts
4	Micas	thin sheet structure black-biotite, white-muscovite	mica schists; some gneiss
5	Limonite	yellow or yellowish-brown; earthy	Iron ores
6	Hematite	red or reddish-brown; earthy	Iron ores
7	Calcite	white, light-gray or colourless	limestone, marble, chalk, marl
8	Dolomite	white, buff, or colourless	dolomite and dolomitic limestone and marble
9	Apatite	brownish or reddish greasy lustre	phosphate beds
10	Gypsum	colourless, white, pink or gray	some salt beds
11	Silicate clay minerals	earthy-white or light-gray	shales and clays

The various minerals in the rocks are the source of inorganic elements on which plants depend for their nutrition in growth and maturity. The alteration of the minerals during the process of weathering gives rise to the formation clay and clay minerals. The elements fixed in chemical combination in the form of rock minerals are not readily available to plants for their nutrition. The process of weathering releases from these minerals the elements required in plant nutrition in forms which are soluble or are readily assimilable by plants. Table VI gives a general picture of the composition of the various minerals and the elements for plant nutrition.

TABLE VI

SOIL MINERALS

Mineral	*Significance in relation to soils* — *Plant nutrient elements contained**	*Significance in relation to soils* — *Other relationships of major importance*
Quartz	None	Major constituent of most sands and silts
Feldspars—Orthoclase	Potassium	Weathering produces clay minerals
Feldspars—Plagioclase	Calcium	
Augite and Hornblende	Calcium, Magnesium, Iron	Weathering produces clay minerals
Micas—Biotite	Potassium, Magnesium, Iron	Weathering produces clay minerals
Micas—Muscovite		
Limonite	Iron	Imparts yellow and yellow brown colours to soils
Hematite	Iron	Imparts red and reddish-brown colours to soils
Calcite	Calcium	Principal constituent of liming material. Occurs as "natural" lime in some soils
Dolomite	Calcium, Magnesium	Principal constituent of liming material. Occurs as 'natural' lime in some soil.
Apatite	Phosphorus, Calcium	Raw material of phosphate fertilizers
Gypsum	Sulphur, Calcium	Treatment of alkali soil. Occurs in subsoils of some dry-land soils
Silicate clay minerals	Absorbed ions of the base	Most active constituent of soil mineral matter (chemically and physically); often colloidal

*Largely "unavailable" to plants until weathering, often very slowly, has taken place.

SOURCE: Max Jensen and Claude H. Pair, "Water," *U.S.D.A. Year Book, 1965*, p. 713.

The formation of soil from the parent rock is the resultant of the action of different forces and the nature of the soil thus formed is influenced by the following factors: nature of parent rock, climate, vegetation, topography, and time. All of these factors individually and collectively control the nature and properties of the soil formed and also the rate of its formation and leave an imprint in the soil profile that is developed in consequence. A study of the soil profile can help to characterise the physical nature and properties of the soil.

It must be emphasised that soil develops from parent material by a process of soil formation that differs from processes of rock weathering, which produces parent material. In the development of parent material the rocks and minerals are broken down into an unconsolidated mass by destructive weathering. As a soil is developed from the parent material, certain changes take place which differentiate the various horizons to give soils their distinctive layering. These changes are brought about largely by shifting of materials either mechanically or in solution to various layers within the horizons or completely out of the soil. The introduction of plant and animal life greatly facilitates the shifting of these materials. The term "eluviation" has been adopted to designate this movement of materials either mechanically or chemically.

THE SOIL PROFILE

If a section downward through a representative soil is examined, a layering, often well defined, will usually be found. Such a section is called a "profile" and the individual layers are called "horizons." The horizons above the parent material are collectively referred to as the "solum." Every well-developed undisturbed soil has its own distinctive profile characteristics. These are made use of in soil survey classification and are of great practical importance. In judging a soil, its whole profile should be taken into consideration unless its depth is far beyond that of normal root penetration.

The upper layers of a soil profile generally contain considerable amounts of organic matter, and are usually darkened appreciably because of such an accumulation. Layers thus characterised are conveniently referred to as the major zone of organic matter accumulation. When ploughed and cultivated, they make up the familiar surface soil or furrow slice.

The underlying subsoil, also markedly weathered, contains comparatively less organic matter. The various subsoil layers, especially in mature humid region soils, often present two very general belts: (1) an upper zone of transition, and (2) a lower zone of accumulation. In the latter, iron and aluminium oxides, clay and even $CaCO_3$ may generally be concentrated. The different layers or horizons that can be found in a soil formed from a

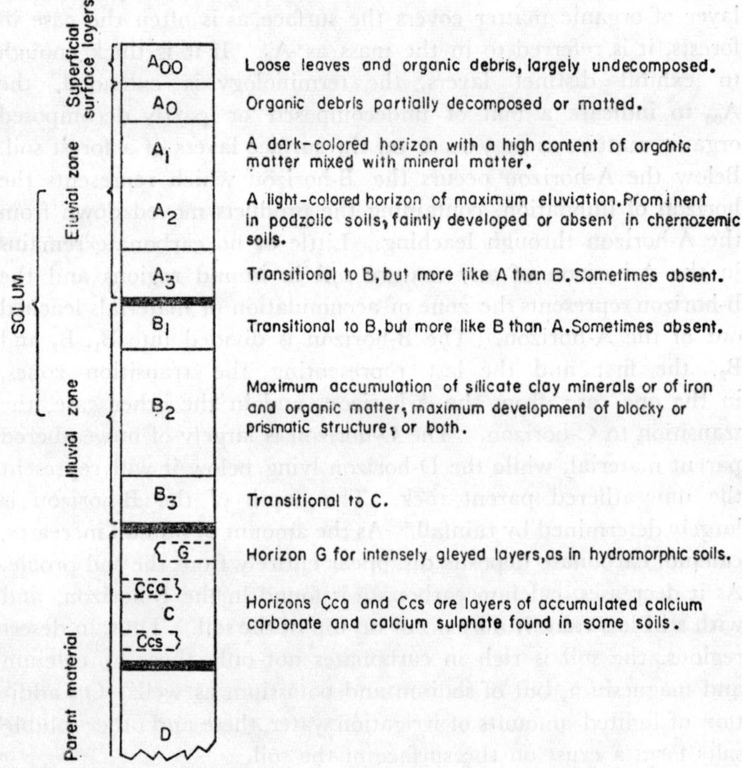

Fig. 16. A diagrammatic sketch showing the arrangements of different horizons in a profile.

parent rock are depicted in the hypothetical soil profile shown in Fig. 16.

SOIL HORIZONS

For convenience in study and description, the layers resulting from soil-forming processes are grouped under three heads: A, B and C. The subdivisions under these are called horizons. The A group lies at the surface, and is characterised as a zone of maximum leaching or eluviation. Beginning at the surface of the mineral matter, the horizons are designated as A_1, A_2, etc. If a layer of organic matter covers the surface, as is often the case in forests, it is referred to in the mass as A_0. If it is thick enough to exhibit distinct layers, the terminology is extended, the A_{00} to indicate a mat of undecomposed or partly decomposed organic matter, as may occur on the surface layers of a forest soil. Below the A-horizon occurs the B-horizon which represents the horizon of illuviation, containing the products moved down from the A-horizon through leaching. Little or no carbonate remains in the A-horizon of any mature soil in humid regions and the B-horizon represents the zone of accumulation of materials leached out of the A-horizon. The B-horizon is divided into B_1, B_2 and B_3, the first and the last representing the transition zones, in the one case, from the A-horizon, and in the other case, the transition to C-horizon. The C-horizon is largely of unweathered parent material, while the D-horizon lying below it will represent the unweathered parent rock. The depth of the B-horizon is largely determined by rainfall. As the amount of rainfall increases, calcium carbonate deposits disappear entirely from the soil profile. As it decreases, calcium carbonate is found in the B-horizon, and with still less rain, it may occur on top of the soil. Thus, in desert regions, the soil is rich in carbonates not only those of calcium and magnesium, but of sodium and potassium as well. On addition of limited amounts of irrigation water, these and other soluble salts form a crust on the surface of the soil.

The development of characteristic differences in soil formation, as recognised by the variations in profile characters, has been made use of in classifying soils and grouping them. Soils are now classified into orders, suborders and great soil groups, series and types. The three major divisions of orders are Zonal, Intrazonal and Azonal. The Zonal order includes soils with well-defined characteristics that reflect the influence of climate and

FIG. 17. Detailed study of the morphological characteristics of soil profile is essential for describing the soils.

vegetation. The Intrazonal order contains soils that reflect dominance of some local factors such as topography, parent material, or age over the influence of climate and vegetation. The Azonal order embraces all the remaining soils that have no well-developed characteristics. Some 36 great soil groups are recognised, which are classified under different suborders and orders.

Each of the great soil groups is divided into series. All the members of a soil series are soils which show similarity in their formations, soils with similar profile characters being grouped in the same series. In addition to a common mode of origin, similarity in topography and drainage, in range of depth, colour, structure, and the reaction of their A and B-horizons characterise the soils belonging to the same series. The series is named usually after the locality in which the soil was first recognised and mapped separately.

The final unit of classification is the soil type. Series are divided into types on the basis of texture of the top soil. A large number of types of the same series is possible, but it is seldom that all possible types are found in any one series.

The various soil orders, suborders and groups of soils of the world are listed in Table VII.

TABLE VII

THE GREAT SOIL GROUPS OF THE WORLD (BALDWIN, *et al.*)*

Order 1	*Suborder* 2	*Great soil groups* 3
	Soils of the Cold Zone	1 Tundra soils
	1 Light-coloured soils of arid regions	2 Desert soils
		3 Red desert soils
		4 Sierozem
		5 Brown soils
		6 Reddish-brown soils
Zonal soils	2 Dark-coloured soils of the semi-arid sub-humid and humid grasslands	7 Chestnut soils
		8 Reddish-chestnut soils
		9 Chernozem soils
		10 Prairie soils
		11 Reddish Prairie soils
	3 Soils of the forest grassland transition	12 Degraded Chernozem soils
		13 Non-calcic brown soils

*Baldwin M., Kellogg C. E., and Thorp. J. "Soil classification," *U. S. D. A. Year Book, 1938.*

Table VII (*continued*)

1	2	3
	4 Light-coloured podsolic soils of forested warm temperate and tropical regions	14 Podsol soils 15 Brown podsolic soils 16 Gray-brown podsolic soils
	5 Laterite soils of forested warm temperate and tropical regions	17 Yellow podsolic soils 18 Red podsolic soils 19 Yellow-brown lateritic soils 20 Reddish-brown lateritic soils 21 Laterite soils
Intrazonal soils	1 Halomorphic (Saline and alkali soils)	1 Solonchak or saline soils 2 Solonetz soils 3 Soloth (degraded alkali) soils
	2 Hydromorphic soils of swamp marshes, etc.	4 Wiesenboden 5 Alpine meadow soils 6 Bog soils 7 Half-bog soils 8 Planosols 9 Ground water podsol soils 10 Ground water laterite soils
	3 Calcimorphic	11 Brown forest soils 12 Rendzina soils
Azonal soils		1 Lithosols 2 Alluvial soils 3 Sands (dry)

Counterparts of many of these orders and suborders are found in India. Though the nomenclature adopted may not be identical in all cases, but according to their characteristics they can be correlated with the corresponding great soil groups. The classificational system of Baldwin *et al.*, was described as a classification according to soil characteristics, but it was no more so than many others that preceded it. The Azonal order was defined in terms of soil characteristics, while the zonal and intrazonal orders, however, were defined provisionally in genetic terms. Several major defects have been common to nearly every system proposed. The great soil groups were defined in the American system in terms of soil properties, but the definitions were brief and serious differences of opinion developed on the interpretation of the definitions. To be useful, definitions must be precise.

Taking note of various lacunae in the soil classification systems proposed earlier, the soil survey staff of the U.S.D.A. set out an elaborate programme to draw up a new and natural classification of the soils of the United States, which could accommodate soils of other countries also. This system after various trials in the field was introduced in 1960 as the "7th approximation of a comprehensive system of soil classification." This provided for 10 orders with over 29 suborders and over 100 great groups.

These orders can perhaps be best introduced by relating them to the kinds of soil recognised in previous classifications. An attempt is made in the table below to relate the soil orders according to 7th Approximation to the soils according to the revised classification of Baldwin et al.

TABLE VIII

FORMATIVE ELEMENTS IN NAME OF SOIL ORDERS

No. of order	*Name of order*	*Formative element in name of order*	*Derivation of formative element*	*Mnemonicon and pronunciation of formative elements*
1	Entisol	ent	Nonsense syllable	recent
2	Vertisol	ert	L. *verto*, turn	invert
3	Inceptisol	ept	L. *inceptum*, beginning	inception
4	Aridisol	id	L. *aridus*, dry	arid
5	Mollisol	oll	L. *mollis*, soft	mollity
6	Spodosol	od	GK. *spodos*, wood ash	Podzol; odd
7	Alfisol	alf	Nonsense syllable	Pedalfer
8	Ultisol	ult	L. *ultimus*, last	Ultimate
9	Oxisol	ox	F. *oxide*, oxide	oxide
10	Histosol	ist	G. *histos*, tissue	histology

In Table X are set out the formative elements and the derivation of these formative elements as used in the names of suborders, of great groups of sub-groups, which help to throw light on how this nomenclature is to be used in naming soil group according to this system.

TABLE IX

	Orders	Approximate equivalents
1	Entisols	Azonal soils, and some Low Humic Gley soils
2	Vertisols	Grumusols
3	Inceptisols	Ando, Sol Brune Acids, some Brown Forest, Low Humic-Gley, and Humic Gley soils
4	Aridisols	Desert, Reddish-Desert, Sierozem, Solonchak, some Brown and Reddish-Brown soils, and associated Solonetz
5	Mollisols	Chestnut, Chernozem, Brunizem (Prairie), Rendzinas, some Brown Forest, and associated Solonetz and Humic Gley soils
6	Spodosols	Podzols, Brown Podzolic soils, and Groundwater Podzols
7	Alfisols	Grey-Brown Podzolic, Grey-Wooded soils, Noncalcic Brown soils, Degraded Chernozem, and associated Planosols and some Half Bog soils
8	Ultisols	Red-Yellow Podzolic soils, Reddish-Brown Lateritic soils of the U.S., and associated planosols and Half Bog soils
9	Oxisols	Laterite soils, Latosols
10	Histosols	Bog soils

The names given to these are combinations of Greek, Latin and English words and are intended to signify various factors of genetic influence. This system has, undergone various changes since 1960, and an elaboration of this system was brought out as a Supplement to the 7th Approximation during 1967, and this Supplement showed many changes in the description for suborders and great groups given earlier, besides introducing some new ones.

In the new system of classification, the ten orders are recognised based on diagnostic horizons, degree of horizonation, presence or absence of certain horizons, and gross composition. The intent has been to choose as differentiating characteristics, properties

Table X

FORMATIVE ELEMENTS IN NAMES OF SUBORDERS

Formative elements	*Derivation of formative element*	*Mnemonicon*	*Connotation of formative element*
alb	L. *albus*, white	albino	Presence of albic horizon (a bleached eluvial horizon)
and	Modified from *Ando*.	Ando	Ando-like
aqu	L. *aqua*, water.	aquarium	Characteristics associated with wetness
ar	L. *arare*, to plow	arable	Mixed horizons
arg	Modified from argillic horizon; L. *argilla*, white clay	argillite	Presence of argillic horizon (a horizon with alluvial clay)
bor	Gr. *boreas*, northern	boreal	Cool
ferr	L. *ferrum*, iron	ferruginous	Presence of iron
fibr	L. *fibra*, fiber	fibrous	Least decomposed stage
fluv	L. *fluvius*, river	fluvial	Flood plains
hem	Gr. *hemi*, half	hemisphere	Intermediate stage of decomposition
hum	L. *humus*, earth	humus	Presence of organic matter
lept	Gr. *leptos*, thin	leptometer	Thin horizon
ochr	Gr. base of *ochros*, pale	ocher	Presence of ochric epipedon (a light-coloured surface)
orth	Gr. *orthos*, ture	orthophonic	The common ones
plag	Modified from Ger. *plaggen*, sod		Presence of plaggen epipedon
psamm	Gr. *psammos*, sand	psammite	sand textures
rend	Modified from Rendzina	rendzina	Rendzina-like
sapr	Gr. *sapros*, rotten	saprophyte	Most decomposed stage
torr	L. *torridus*, hot and dry	torrid	Usually dry
trop	Modified from Gr. *tropikos*, of the solstice	tropical	Continually warm
ud	L. *udus*, humid	udometer	Of humid climates
umbr	L. *umbra*, shade	umbrella	Presence of umbric epipedon (a dark-coloured surface)
ust	L. *ustus*, burnt	combustion	Of dry climates, usually hot in summer
xer	Gr. *xeros*, dry	xerophyte	Annual dry season

The names of these orders are given in Table XII together with the formative elements in the name of the order and the derivative of these formative elements.

NAMES OF SUBGROUPS

Subgroup names consist of the name of the appropriate great group modified by one or more adjectives. The adjective *typic* is used for the subgroup that is thought to typify the central concept of the great group.

The name of subgroups that we might call intergrades or ectragrades, together with their derivations are as follows:

TABLE XI

Formative element	Derivation of formative element	Mnemonicon	Connotation of formative element
abruptic	L. *abruptum*, torn off	abrupt	Abrupt textural change
allic	Modified from *aluminum*		Presence of extractable aluminium
arenic	L. *arena*, sand	arenose	Sandy texture
clastic	Gr. *klastos*, broken	clastic	High mineral content
cumulis	L. *cumulus*, heap	accumulation	Thickened epipedon
glossic	Gr. *glossa*, tongue	glossary	Tongued
glossarenic	L. *grossus*, thick, and L. *arena*, sand		Thick sandy layer
limnic	Modified from Gr. *limn*, lake	limnology	Presence of a limnic layer
lithic	Gr. *lithos*, stone	lithosphere	Presence of a lithic contact
leptic	Gr. *leptos*. thin		A thin solum
pergelic	L. *per*, throughout in time and space, and L. *gelare*, to freeze		Permanently frozen or having permafrost
petrocalcic	Gr. *petra*, rock and *calcic* from calcium		Petrocalcic horizon
plinthic	Modified from Gr. *plinthos*, brick	plinthite	Presence of plinthite
ruptic	L. *ruptum*, broken	rupture	Intermittent or broken horizons
stratic	L. *stratum*, a covering	stratified	Stratified layers
superic	L. *superare*, to overtop	superimpose	Presence of plinthite in the surface
pachic	Gr. *pachys* thick	pachyderm	A thick epipedon

TABLE XII

FORMATIVE ELEMENTS FOR NAMES OF GREAT CROPS

Formative element	Derivation of formative element	Mnemonicon	Connotation of formative element
acr	Modified from Gr. *akros*, at the end	acrolith	Extreme weathering
agr	L. *ager*, field	agriculture	An agric horizon
alb	L. *albus*, white	albino	An albic horizon
and	Modified from *Ando*	Ando	Ando-like
anthr	Gr. *anthropos*, man	anthropology	An anthropic epipedon
aqu	L. *aqua*, water	aquarium	Characteristic associated with wetness
arg	Modified from argillic horizon; L. *argilla*, white clay	argillite	An agrillic horizon
calc	L. calcis, *lime*	calcium	A calacic horizon
camb	L. L. *cambiare*, to exchange	change	A cambic horizon
chrom	Gr. *chroma*, colour	chroma	High chroma
cry	Gr. *kryos*, coldness	crystal	Cold
dur	L. *durus*, hard	durable	A duripan
dystr, dys	Modified from Gr. dys, ill; *dystrophic*, infertile	dystrophic	Low base saturation
eutr, eu	Modified from Gr. *eu* good; *eutrophic* fertile	eutrophic	High base saturation
ferr	L. *ferrum*, iron	ferric	Presence of iron
frag	Modified from L. *fragilis*, brittle	fragile	Presence of fragipan
fragloss	Compound of *fra*(*g*) and *gloss*		See the formative elements *frag* and *gloss*
gibbs	Modified from *gibbsite*	gibbsite	Presence of gibbsite
gloss	Gr. *glossa*, tongue	glossary	Tongued
hal	Gr. *hals*, salt	halophyte	Salty
hapl	Gr. *haplous*, simple	haploid	Minimum horizon
hum	L. *humus*, earth		Presence of humus
hydr	Gr. *hydor*, water	hydrophobia	Presence of water
hyp	Gr. *hypnon*, moss	hypnum	Presence of hypnum moss

Table XII (*continued*)

luo, lu	Gr. *louo*, to wash	ablution	Illuvial
moll	L. *mollis*, soft	molligy	Presence of mollic epipedon
nadur	Compound of *na*(*tr*), and *dur*		
natr	Modified from *natrium*, sodium		Presence of natric horizon
ochr	Gr. base of *ochros*, pale	ocher	Presence of ochric epipedon (a light-coloured surface)
pale	Gr. *paleos*, old	paleosol	Old development
pell	Gr. *pellos*, dusky		Low chroma
plac	Gr. base of *plax*, flat stone		Presence of a thin pan
plag	Modified from Ger. *plagern*, sod		Presence of plaggen horizon
plinth	Gr. *plinthos*; brick		Presence of plinthite
quartz	Ger. *quarz*, quartz	quartz	High quartz content
rend	Modified from rendzina	rendzina	Rendzina-like
rhod	Gr. base of *rhodon*, rose	rhododendron	Dark-red colours
sal	L. base of *sal*, salt	saline	Presence of salic horizon
sider	Gr. *sideros*, iron	siderite	Presence of free iron oxides
sphagno	Gr. *sphagnos*, bog	sphagnum-moss	Presence of sphagnum-moss
torr	L. *torridus*, hot and dry	torrid	Usually dry
trop	Modified from Gr. *tropikos*, of the solstice	tropical	Continually warm
ud	L. *udus*, humid	udometer	Of humid climates
umbr	L. base of *umbra*, shade	umbrella	Presence of umbric epipedon
ust	L. base of *ustus*, burnt	combustion	Dry climate, usually hot in summer
verm	L. base of *vermes*, worm	vermiform	Wormy, or mixed by animals
vitr	L. *vitrum*, glass	vitreous	Presence of glass
xer	Gr. *xeros*, dry	xerophyte	Annual dry season
sombr	F. *sombre*, dark	somber	A dark horizon

that reflect major differences in the genesis of soils. Moisture regime, temperature, mineralogy, and specific kinds of horizons are considered in differentiating suborders within the orders. Great Groups are distinguished within suborders by the presence or absence of characteristic horizons or other features. The range of definitive characters at Great Group level have been narrowed down to make the soils more homogeneous in their characteristics than the soils of classes in higher categories.

The classification at the family level, through subgroup is mainly based on soil temperature, texture, and mineralogy, which have a direct bearing on plant growth. The new system of classification and the sequential nomenclature make the characteristics of soils easier to understand and remember. This also brings out relationship among soils and between the soil and environment, and provides a basis for developing principles of soil genesis and soil behaviour that have prediction value.

REFERENCES

Jenny, *Factors of Soil Formation*, McGraw Hill Book Co., New York, 1941.

Leefer, G.W., *Introduction to Soil Science*, Cambridge University Press, 1952.

Kellogg, C.E., *The Soil That Supports Us*, Macmillan & Co., New York, 1947.

Miller, C.E., and Turk, L.M., *Fundamentals of Soil Science*, John Wiley & Sons, New York, 1967.

Soil Survey staff, S.C.S.U.S.D.A., *Supplement to the Soil Classification System, 7th Approximation*, 1967.

CHAPTER V

SOILS OF INDIA — OUR HERITAGE

GEOLOGICAL FORMATIONS AND CLIMATIC RANGES

THE soils of India are derived from a wide variety of parent rocks and materials. Their composition varies over a wide range both in the nature and kind of minerals present in them and also in their proportions. The origin of these parent rocks is varied and they have been formed over different geological periods. Of these, the peninsular part of India, lying to the south of the Vindhya range of hills, has an older geological history compared with the rest of the country. Over the Indian subcontinent, three main geological regions may be recognised, viz. the Peninsular region, the Indo-Gangetic plains and the Himalayan region. The Peninsular region comprises the area located to the south of the Vindhyas and bordered by the range of hills along the western and eastern limits of the Peninsula which end at Cape Comorin at the southernmost tip of the Peninsula. The Himalayan region at the northern limits of the subcontinent represents the highly folded sedimentary rocks running from the west in the Punjab and Himachal Pradesh towards north in Kashmir and Ladakh, and then turning east in a south-eastern direction through Uttar Pradesh and ending in Assam. The third vast region, viz. the great Indo-Gangetic plain, representing the largest extent of most intensively populated area in the world, nestles between the two regions mentioned above.

The rocks of the Peninsular region are ancient archaean formations which have never been under the sea and, therefore, are free from massive deposits or extensive sedimentary formations. These areas have been exposed to minimum of thrusts and stresses to which the extra peninsular formations have been subjected to during different geological epochs. The rocks of the Himalayan formations are of comparatively recent geological origin and have been formed by the upthrust of the earth's crust caused by powerful forces operating from the northern region, with the Peninsular region offering resistance and setting up forces opposing the ones from the north. Geologically, the Himalayan region has been divided into three zones as follows:

(i) The Northern or the Tibetan Zone: This zone lies behind the line of the highest elevations, extending beyond the elevation of 11,000 feet, and is composed of a series of highly fossiliferous marine sedimentary rocks.

(ii) The Central Zone, which represents most of the central and middle regions, is composed of crystalline and metamorphic rocks like granites, gneisses and schists.

(iii) The outer or sub-Himalayan Zone, which consists of the Siwalik formations and is formed of sedimentary river deposits.

The vast Indo-Gangetic plain lies between Punjab in the west and West Bengal in the east. It has been formed by the deposits of silt and clay brought down from the mountainous region of the Himalayas by the river systems of the north. The thickness of the alluvial deposits range from 100 to 10,000 ft, with many areas possessing thickness far in excess of the latter figure.

With regard to the forces of climate which are important factors in the development of soil, the range of these forces varies over wide extremes, and is related to the variation in the climate in different parts of the country. India lies partly in the subtropics and partly in the tropics. The line of Cancer passes through the Peninsular region. The temperature variations range from the sub-zero temperatures of Ladakh to the burning tropical heat of 120°F in parts of the desert region of the north-west. The rainfall and wind velocities show variations which are as wide as the range in temperatures. The rainfall is as low as 4″ per annum, as it is over a great part of western Rajasthan, and as high as 450″ in parts of Assam and in the Western Ghats in Mysore. Over most parts of the country, two important seasons are recognised when monsoon rains fall, spreading over a period of time: South-West Monsoon (June to September) and North-East Monsoon (November and December).

The wind velocities vary widely during the different periods of the year and over different parts of the country and range from a still breeze of 2 to 4 mph to the cyclonic winds which may whip up to velocities of 80 to 100 mph and above.

THE SOIL MAP OF INDIA

With the wide range of geological materials forming the surface of the subcontinent and a similarly wide range of climatic factors and their associated factors of vegetation acting on them, it is to be expected that a wide variety of soils is developed, on the land surface. Classification of the soils occurring in India has been attempted over a period of years, and soil maps of India have been prepared during the past three decades, and these have successively undergone various changes, modifications and improvements, as more accurate and more reliable information became available. The earliest soil map of India was drawn up in 1932 by Sokolasky, followed by a map by Vishwanath and Ukil in 1944. This map was further improved upon in 1952 by Raychaudhury. Based upon more recent information collected from the various States and the work of the All India Soil Survey, Govinda Rajan has compiled a revised soil map of India which indicates the distribution of 23 different soils. A map of the country showing the distribution of these different soils is given in Fig. 18. A brief description of the characteristics of each of the broad groups of soils represented on the map is given in the following paragraph.

RED SOILS

The red soils or the red earths occur in India mainly in the Peninsular portion in the South and along the east coast going up to Assam in the north-east. Parts of east-central India also have the red soils. These soils vary considerably in texture, some being sandy, while a great many are loams of different kinds and a few are clay soils. The two predominant types that have been distinguished for purposes of delineation in the Soil Map of India are the red loams and the red sandy soils, and this differentiation also seems to be related to their genesis from geological materials of distinct compositional differences. In general, the red soils as they occur in Peninsular India are derived from granites and gneisses, mainly of the Archaean period. The other rock formations from which red soils are developed are sandstones, hornblende and mica schists, acid traps, quartzites and shales.

RED LOAM SOILS

The red-loam soils in particular are formed by the weathering of the rocks like gneisses, charnockites, diorites and others which are relatively richer in the clay-forming minerals and correspondingly poorer in the acid component, viz. silica. The weathering of feldspar, mica, hornblende or other base rich minerals results in the formation of fine-textured soils wherever free silica is relatively less abundant. In some red soils, lime concretions in the form of nodules or thick veins are found. The latter result from the weathering of the feldspar containing lime in the rocks which occur as intrusions or as massive veins. The texture of these soils may vary from loam to silty clay and clay loam. The normal red loam soils have a pH around neutrality, or else slightly on the acid side, while the soils containing lime may show a pH as low as 8.0. The A-horizon which may have a depth ranging from 25 to 40 cms is a dark reddish-brown (5YR 3/2) loam with a crumb structure, and the B-horizon going down to 70 to 80 cms from the surface may have a dark-brown (7.5 YR 3/2) loam or clay loam with a weak angular blocky structure. Below this layer is met soft disintegrating weathered gneissic rock and, in the case of lime-bearing rocks there may be thick veins or beddings of soft and hard calcareous deposits resulting from the weathering of the lime rich feldspathic gneiss. In such lime-bearing soils, the B-horizon also may have lime concretions of varying sizes distributed in them.

RED SANDY SOILS

These soils are generally derived from granites, coarse grained granites, granatoid gneisses, quartzites, sandstones, etc. and are characterised by being rich in the fine and coarse sand fractions. The clay minerals become coated with red haematite or yellow limonite or a mixture of the two oxides of iron forming a red, yellow or reddish-yellow soil. The yellowish soil becomes red when the limonite undergoes dehydration and changes to haematite. Ferruginous gravel formed of impure iron, alumina and silica concretions and bits of quartz are the common accessory constituents of red soils. The characteristic clay minerals in these soils

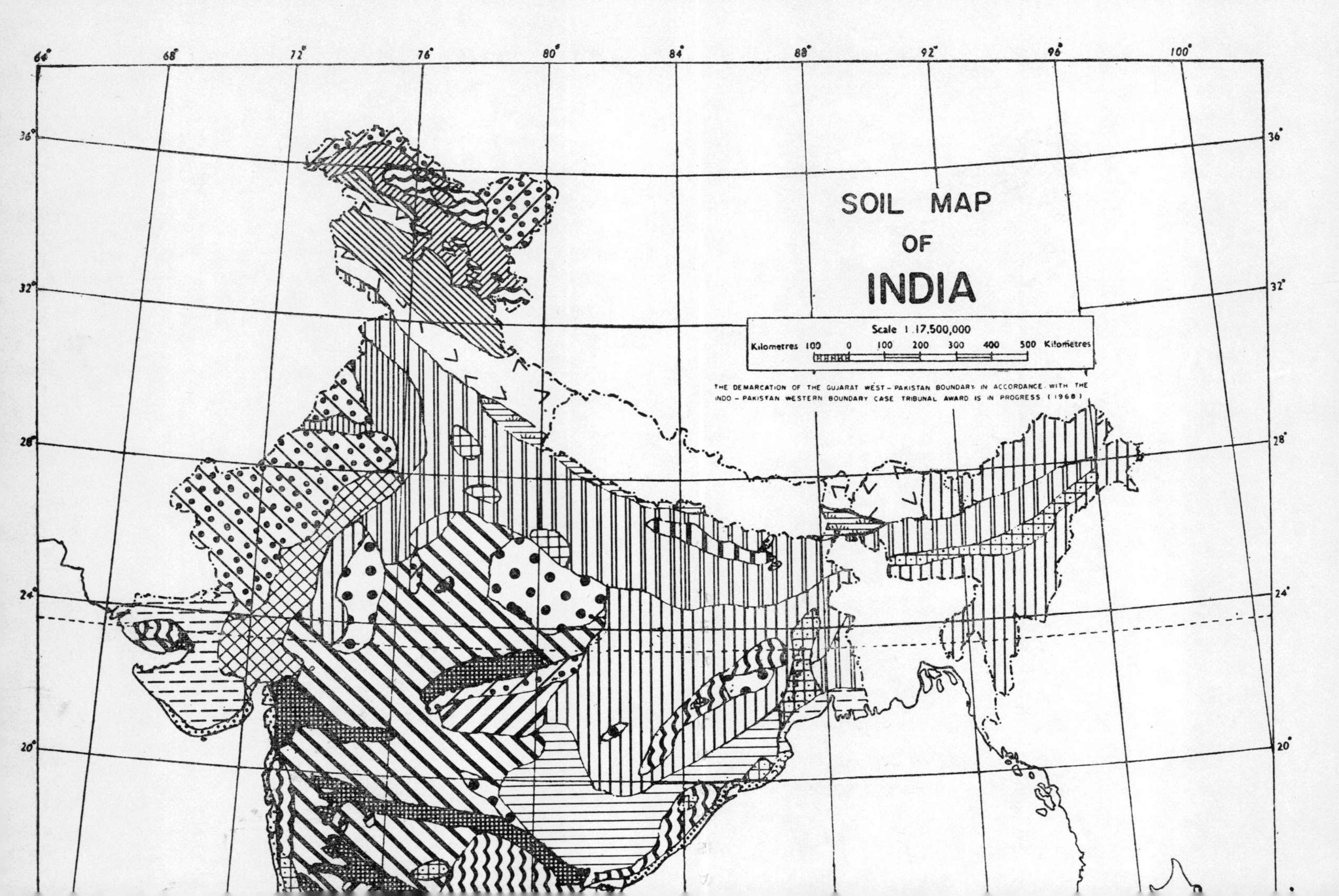
SOIL MAP
OF
INDIA
Scale 1.17,500,000
Kilometres 100 0 100 200 300 400 500 Kilometres
THE DEMARCATION OF THE GUJARAT WEST - PAKISTAN BOUNDARY IN ACCORDANCE WITH THE
INDO - PAKISTAN WESTERN BOUNDARY CASE TRIBUNAL AWARD IS IN PROGRESS (1968)
64°
68°
72°
76°
80°
84°
88°
92°
96°
100°
36°
32°
28°
24°
20°

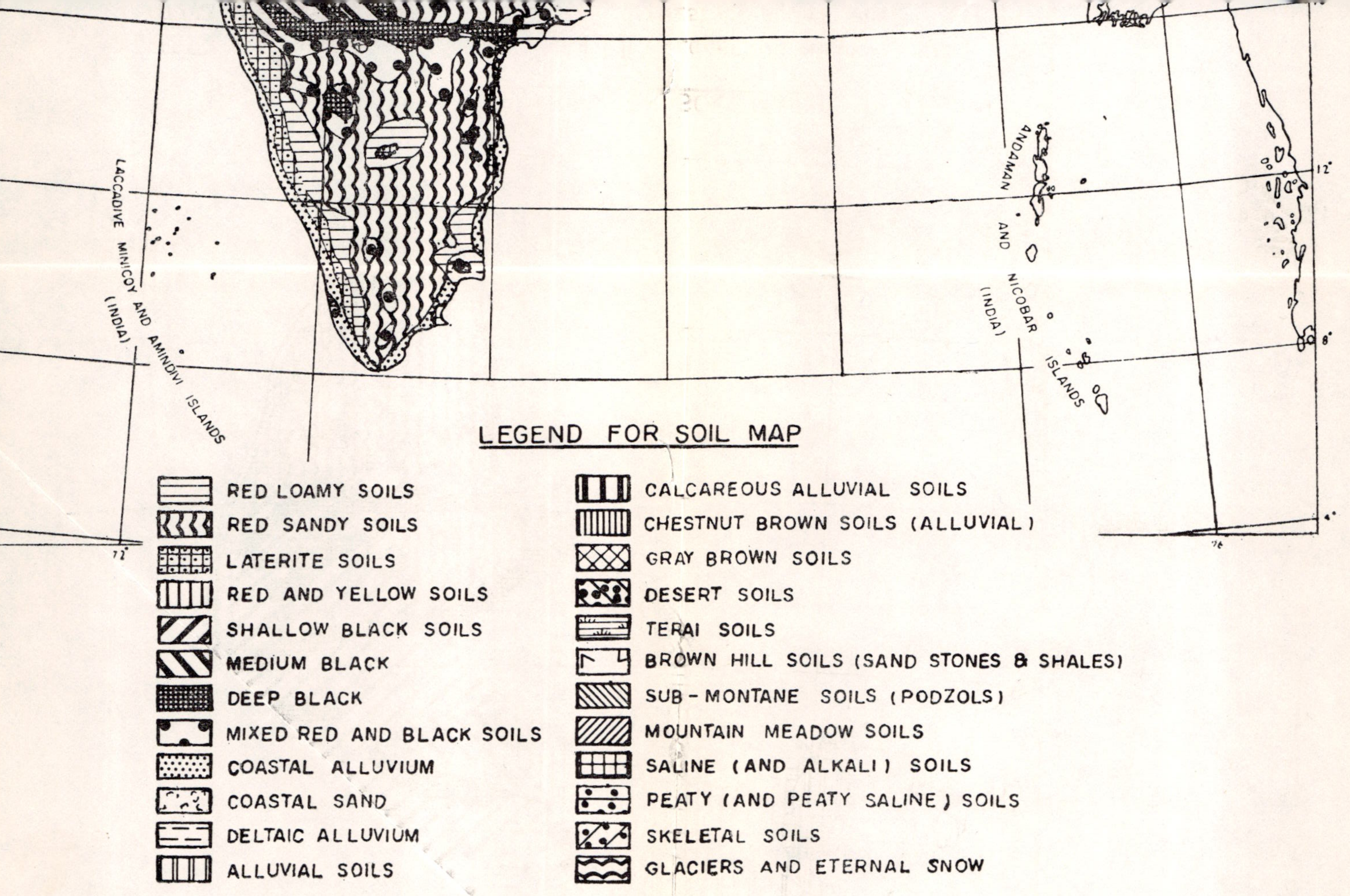

FIG. 18. A soil map of India indicating distribution of soils.

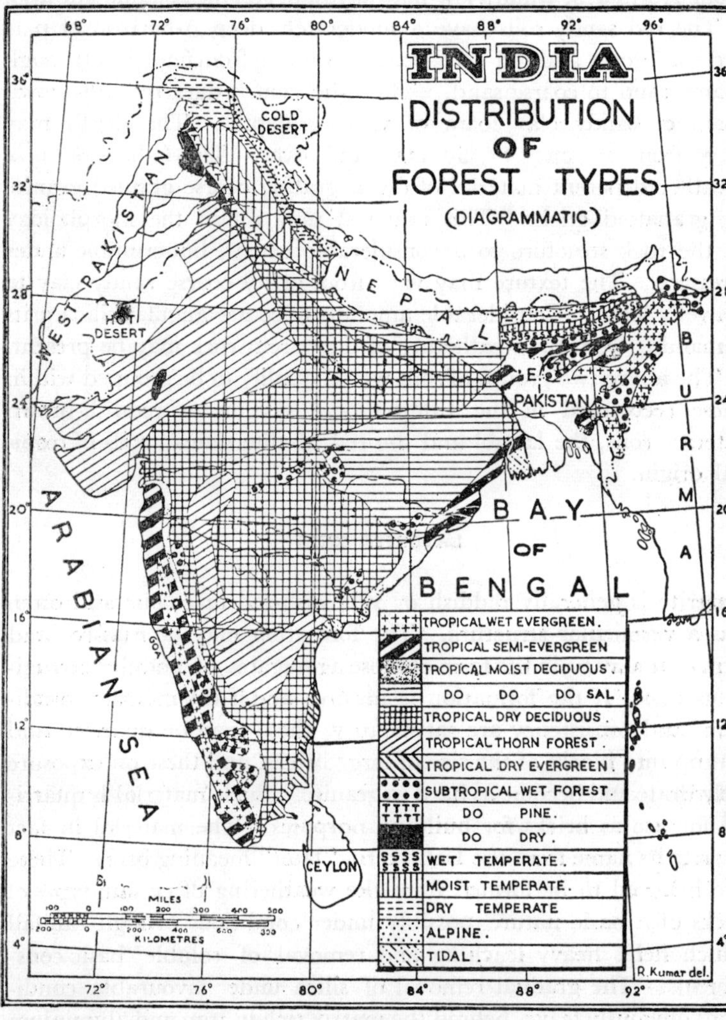

FIG. 19. A map of India showing the distribution of forest types.

are composed mainly of the kaolinitic and illinitic types of minerals with smaller amounts of montmorillonite. The base exchange capacity of the soils, generally varies from 5 to 25 me per 100 grams depending upon the amounts of clay and organic matter. The majority of these soils are slightly on the acid side with pH ranging from 4.5 to 6.5, while a few may have a pH on the alkaline side also.

The red sandy soils have a moderately deep A-horizon of pale reddish brown and red with textures varying from fine sand through loamy sand to coarse sand. The B-horizon is dark reddish-brown loam or sandy clay loam of varying depths. The depth may vary from 15 cms to 30 cms and over. The C-horizon is a weathered parent material of disintegrating coarse grained granite or granatoid gneiss. This material may retain the morphology of the rock structure to a considerable extent, but crumble under pressure. The texture may vary from gritty coarse sandy clay to gravelly loam. Considerable amounts of coarse angular silica grits remaining from the weathering of the parent rock may be present.

The above two groups of soils would qualify to be grouped within these recognised in the International systems as reddish-brown lateritic soils, the latosol and the red Mediterranean soils of tropical origin.

LATERITE SOILS

Laterite is generally reddish or yellowish-red in colour and often has a vermicular structure. The formation may be massive and firm or it may be in the form of loose aggregates of nodular ferruginous mass. If the formation is massive and if the moisture conditions and consistency are satisfactory, these may be quarried and cut up into blocks of the size of large bricks, and these on exposure dehydrate and become as hard as granite. Such material is quarried for use as bricks for building purposes. The material in fact derives its name from the Latin term "later" meaning brick. These are believed to be formed from the weathering of certain types of rocks of a basic nature possibly under conditions of high rainfall which helps heavy leaching and removal of soluble basic constituents. The gradual removal of silica under favourable conditions obviously leaves behind the matrix rich in iron and aluminium oxides and hydroxides and some silica.

Lateritic soils are those which are associated with and derived from laterites. The decomposition of only a part of the colloidal complex is presumed to have taken place and there is a larger proportion of the primary kaolinitic minerals than in laterites. As the laterites do not contain primary clay minerals, they show the typical properties of clay such as plasticity, cohesion, shrinkage, expansion, base exchange properties, etc. only to a small extent. The base exchange capacity of the mineral colloids may range from 2 to 4 me

FIG. 20. Laterite bricks quarried from a laterite area are kept out in the sun for drying. This makes the bricks hard and fit for construction work.

per 100 grams for laterites and from 4 to 7 for laterite soils. The molecular silica to sesquioxide ratio of laterite soils is below 1 : 35. The soils are in some cases deep but in many areas rather shallow.

RED AND YELLOW SOILS

In the east-central and in the north-east central parts of India, occur soils which are characteristically yellow in colour. The shade of colour may range from reddish-yellow to yellowish-brown, and the soils are usually fine textured. In some cases, these soils

FIG. 21. A close-up view of a laterite piece showing the characteristic vermicular structure in its mass. Dark colour represents the red and reddish brown iron oxide depositions and the light colours, white, pale yellow, orange or buff, represent materials rich in kaolin, hydrated ferric and aluminium oxides.

are associated with soils developed from laterites or overlying laterites. Otherwise, they are derived from a variety of rocks underlying in the profile at depths varying from 100 to 200 cms. Micaceous quartzite schists, phyllites, hornblende, schists and gneisses, are some of the rocks giving rise to these yellow soils. While, in some cases, poor drainage conditions may be the cause of the yellow colour, the yellow colour of the surface soils cannot be due to this and the soils in general are fairly well drained. The soils have a pH around neutrality or else slightly on the acid side, and it is a moot point if these soils can qualify to be called red-yellow podzolic soils after—those described in south-east United States.

The A-horizon of these soils is loam to silty loam in texture and moderately rich in humus. These soils possess a textural B-horizon with colours of high chroma which are usually yellow, but at times red. This horizon has a blocky structure and at times exhibits considerable shrinkage and cracking on drying. Mottlings and pigmentation due to soft nodular iron may be found in this horizon. In lands cultivated for paddy, a gleyed B-horizon may be found at depths varying from 100 to 150 cms.

BLACK SOILS

Soils with a characteristically dark colour ranging from dark-brown to deep black, occur extensively in the central and south-central parts of India. These have been classified for purposes of mapping in a broad manner into different groups based upon the depth of their formation. Soils possessing a depth of 30 cms or less are described as shallow black soils, those having depths ranging from 30 to 100 cms as medium black soils, and those with a depth in excess of 100 cms, going up to 200 cms and over, as deep black soils.

As a group, these soils are heavy textured and their clay content ranges from 40 to 60 per cent. They are plastic and sticky when wet and very hard when dry. The composition of the clays is generally of the montmorillonitic group, and, as a result, shows very strong swelling and severe shrinkage with changing moisture conditions, causing heavy fissuring and deep cracking on drying. Because of the physical movement of the surface soil to the subsurface layers through the fissures, these soils are popularly described as "self mulching" or "self ploughing" soils. These soils could be grouped in the International systems as "Grumosols" or as Vertisols (7th Approximation) depending upon these characteristics.

SHALLOW BLACK SOILS

These are black soils derived from basalts of the Deccan traps.

The soil is usually silty loam to clay in texture, and the surface has a colour ranging from dark-brown to dark yellowish-brown. The structure is granular or else weakly blocky. Lime in the form of fine grains or nodules is usually present and the soil is freely drained. The solum rapidly merges with the disintegrating parent rock of hard basalt.

MEDIUM BLACK SOILS

These are black soils with depths ranging from 30 to 100 cms and developed from a variety of rocks including basaltic traps, Dharwar schists, basic granites, gneisses, hornblende and chlorite schists. The texture ranges from silty clay to clay and it is difficult to notice any increase in the clay content in the lower horizons. The soils are moderately rich in organic matter and are fairly well drained. While these soils contain lime in varying proportions, some of them may have thick layers of calcareous nodules (20 to 30 cms thick) in the C-horizon. Further differentiation between these soils as they occur in peninsular India is possible on the basis of the occurrence of gypsum or its absence in the sub-surface layers. Where gypsum occurs, it is generally found below three feet and is associated with a high concentration of soluble salts, particularly, sodium and magnesium sulphates. The gypsum occurrence which is in pockets of crystalline masses is usually over the lime-bearing strata.

DEEP BLACK SOILS

These are soils with depths in excess of 120 cms and going down to 200 cms and over. These are derived from basaltic traps and represent a large part of the black soils of the Deccan plateau, commonly identified as the "Regur" soils. Because cotton is the important commercial crop grown on these soils, these are also referred to popularly as the "Black Cotton Soils."

The texture of the surface and sub-surface layers is more or less uniform and ranges from silty clay to clay. The percentage of clay may vary from 40 to 60 per cent and over. Distribution of lime in the form of irregular-shaped nodules of varying sizes, may be uniform throughout the profile with heavier accumulations at depths of over

150 cms. The soil reaction is moderately alkaline (pH 8.0 to 8.5). In common with the other black soils, the clay mineral is of the 2:1 lattice structure and possesses high retentivity of moisture. In areas of low rainfall, 35 to 50 cms annually, the practice of "dry farming" operations helps to conserve the moisture and raise a crop successfully. The presence of soluble salts in the sub-surface layers is usually the cause of deterioration in the structure of the soil and creates difficulties in cultivation after introduction of irrigation in areas where this is possible.

MIXED RED AND BLACK SOILS

The occurrence of red and black coloured soils side by side, in areas of varying sizes and unpredictable patterns, is a fairly common feature in transitional areas where either of these soils are found. In view of the difficulty of separating the location of the individual soils in the Soil Map, these soils have been grouped as a category of the mixed soils. In many cases, in an area of predominantly red soils, while the red soils occur in terrain of elevated topographic features, the black soils may be found in the lower topographic levels. The higher moisture regime and the accumulation of organic matter in the lower levels tend to form a darker coloured soil of heavier texture. The greater mobility of the montmorillonitic clay may also be a factor in assisting the accumulation of these black soils in the lower levels. In certain cases, intrusions of basic rocks, in the midst of predominantly granitic and granatoid gneissic areas of red soils with the kaolinitic type of clay, have resulted in formation of patches of black soil. Basic rocks rich in soda lime, feldspar and a variety of schistose materials give rise to formation of black soils in such conditions. In these cases, it is a matter of common observation that the black soils occur on higher topographic features in relation to the red soils.

Conversely, in areas predominating in black soils, occurrence of patches of red soil is not unusual. In such cases, field observations indicate that the black soils overlie the red soils, the former having been formed from a variety of parent rocks of different composition from those of the red soils. Flows of basalts, trap and even extensive intrusions of basic rocks are found to give rise to such a feature. The black soils being composed of heavy clay are naturally more impervious to drainage and consequently more erosive and these invariably spread over the adjoining red soils and produce, in the transitional zones, brown soils of varying shades, which are best described as a mixture of red and black soils.

ALLUVIAL SOILS

These soils are a very large group, formed by transportation in streams and rivers and deposited over flood plains or along coastal belts. The character of these soils varies a great deal and reflects the nature of soils that occur in the region of their transportation. Their colour, texture, and quality, and their development in regard to profile differentiation vary accordingly. Most of the fresh alluvia show little or no horizonation, while the older alluvium shows distinct profile development. Their colour varies from pale-grey and pale-yellow to deep-black and their texture can vary from coarse sand to heavy clay. In view of the extremes of variation that are possible in these soils and the large variation in their agricultural potential, it has been found convenient to divide this large group of soils into separate groups to distinguish these characteristics.

COASTAL ALLUVIUM

Soils of the coastal belt all along the peninsular region, and extending over varying widths between the sea and the range of hills along the east and the west coasts, are characterised by being relatively recent deposits of alluvial origin. The texture of these soils is extremely variable and range from sandy to silty clay. Except in the older deposits, these soils do not show a prominent horizonal differentiation. The soils are usually deep and the colours range from bright reddish-brown and yellowish-brown to grey and dark-grey. The soils are fertile and occur in the belt of monsoon rainfall. Advantage of both the south-east and north-west monsoon rains, which precipitate between June and November, is taken and paddy is cultivated intensively in these areas.

The composition and mineralogy of these soils is greatly influenced by the parent materials of the catchment areas. Alluvial soils derived from calcareous materials are usually composed of dark-coloured heavy clay, while in areas where the red soils from the granitic and granite gneissic rocks are dominant, the alluvial soils are poorer in fertility, often medium or light-textured and relatively rich in the kaolinitic clays. Frequently, these soils in areas which are not much above sea-level get inundated by sea-water and suffer from the problems of salinity.

COASTAL SAND

Characteristically, in certain areas along the coast, for varying widths, in the Peninsular region occur soils which are sandy, deep and lacking in any profile or soil development. The lack of profile development is attributed to the coarse nature of the parent material. These sandy stretches can be considered to be Regosols. The topography of the sandy Regosols varies from flat to gently undulating with occasional dunes. Salinity is rarely a problem in these areas because of the relatively low water-table and the free drainage. Some areas in the low-lying flat lands can be marshy and saline, in which case, the swampy condition makes the areas unfit for any useful cultivation. These sandy stretches, if the subsoil water-level is not too deep, are used for raising fuel trees, particularly, casuarina, and coconut. When cultivated, the areas of old red sands, particularly, if they contain some clay, are used for raising millets and also seedlings of tobacco and vegetables which are transported and grown on heavier textured soils in nearby areas.

DELTAIC ALLUVIUM

The soils of the deltaic alluvium represent the heterogeneous types of sediments brought by rivers and deposited at their mouths. The east coast of peninsular India is characterised by the formation of deltas at the mouths of the major rivers of the subcontinent that flow into the sea. Of these rivers, the Ganges and the Brahmaputra flow from the Himalayas and drain the extensive plains of northern India; the Mahanadi, the Godavari and Krishna, in the Deccan plateau and the Central India plains, and the Cauvery in the southern part of the peninsula. These rivers carrying the alluvium of the extensive areas they traverse, deposit it at the regions they join the sea, and these deposits form the alluvial soils of these deltas.

The materials from which alluvial soils of the different deltas are formed vary considerably both in composition and in texture. In the great flood plains, the material is mainly clayey, sometimes loamy and sporadically sandy. The Gangetic alluvia reflect the characteristics of the region they flow through, viz. the great alluvial plains of North India. The soils are light-coloured, silty and silty clay. The Mahanadi delta alluvium again is light-brown and light-yellow in colour, and in texture silty loam, loam and sandy loam. These reflect, to a large extent, the colour and texture of the soils of the east-central plateau of Madhya Pradesh and Orissa through which this river and its tributaries traverse. The Godavari and Krishna rivers, draining as they do the black soils derived from the basalt traps, are largely composed of dark-coloured fine-textured silty clays and clays. They also reflect the base rich characteristics of the rock material which have given rise to the black soils of the Deccan plateau. The alluvium of the Cauvery delta, while not truly representing the red sandy and coarse-textured soils of the granitic area through which most of the rivers' courses lie, contains a high proportion of dark-coloured silt and silty clay soils.

The drainage conditions of most of the deltaic soils, other than the Gangetic alluvium, are satisfactory and hence show rarely grey or mottled horizons at the deeper layers. The accumulation of organic matter in the A-horizon, too, varies a great deal and reflects the drainage and cultivation conditions of the respective areas. Thus, the Gangetic alluvium at the mouth of the Ganges, which is largely swampy shows considerable accumulation of organic matter, the natural vegetation being mangrove and such species, while the other alluvium, which is extensively cultivated and cropped, shows much less accumulation of organic matter.

ALLUVIAL SOILS

These soils represent the vast tracts of riverine alluvium of the Indo-Gangetic plain. The deposits of the great rivers, the Jumna, the Ganges and their tributaries, the Gandak, Gomti, Ghagra and others which flow out of the Himalayan ranges, have accumulated over long periods in the northern part of the subcontinent. The area stretches over a length of nearly 1,000 miles in an east to west direction and a width of about 200 miles, characterised by a topography which is monotonously level, with a general gradiant of about 1 per cent from the north-west to the south-east. The depth of the alluvium is great and may extend to many hundreds of metres, though in some places it is shallow. These water-deposited sediments are in a large measure old, though newer deposits are continuously being added, particularly in the areas subjacent to the river courses and those which are subject to periodic flooding, as along the banks of the Damodar and Kosi which are notoriously migratory.

The colour of the soils range from pale-grey, yellow and yellow-brown to dark grey. The texture is generally silty, though loams and silty clay loam soils are not infrequently met with. A well-formed B-horizon possessing strong angular blocky structure is met with, but in areas where accumulation of soluble salt has taken place this may not be prominent. Deposition of lime in the B_3-horizon which might result in strong cementation is frequently met with. These soils, if not badly affected by salts or suffering from bad drainage are fertile and respond well to manuring. Wheat, gram and oil-seed crops are commonly grown, while paddy also is grown in the eastern parts of the region.

CALCAREOUS ALLUVIAL SOILS

These are alluvial soils which occur characteristically along the north-eastern districts of Uttar Pradesh and extending to the north-western parts of Bihar. These are calcareous soils developed on the alluvium brought by the river Gandak, flowing from the Himalayas in a north-west to south-east direction towards the Ganges. The main characteristic of the soil is the high content of $CaCO_3$ (10 to 40 per cent) which is distributed throughout the depth of the profile. The soils are light-coloured, being pale-brown and yellow-brown, and lack in horizonal differentiation. Their texture varies from sandy loam to loam. The pH of the soils is on the alkaline side and the contents of available phosphoric acid and potash are low. The soils are azonal in character.

CHESTNUT BROWN SOILS

These soils also come under the broad group of alluvial soils. The old Indo-Gangetic alluvium of the north-west region of India has been subject to varying climatic conditions, ranging from humid to arid as the distance from the Himalayas increases. Thus, the State of Punjab, where this region lies, has been divided into six climatic zones representing the extremes mentioned above. The soils within the sub-humid climatic zone have attained profile characteristics which may be referred to as the chestnut-brown soils. These soils, in common with the other alluvial soils of this area, are deep and clayey, with clear evidence of mechanical illuviation in the middle horizons. There is a clear indication of the movement of $CaCO_3$ throughout the profile, and a layer of $CaCO_3$ accumulation is reached below 150 cms. The pH of the soil is neutral to slightly alkaline, and the predominant clay mineral is of the 2 :1 lattice type. The soils are deficient in phosphate, with a medium status in organic matter and nitrogen.

GRAY BROWN SOILS

These soils have developed under semi-arid conditions. The texture of the soils is usually light, being sandy loam with a preponderance of coarse sand. There is evidence of mechanical eluviation, the B-horizons being distinctly heavier textured. $CaCO_3$ is present throughout the depth of the profile, with its layeration commonly found fairly near the surface (60 to 90 cms). The predominant base in the exchange complex is Ca. The soils are neutral to alkaline in reaction and the clay mineral is predominantly montmorillonite. The soils are generally poor in nitrogen and phosphorus, but are adequate in potash content.

DESERT SOILS

The soils found characteristically in the arid areas in the north-western region in the states of Rajasthan and the Punjab, and lying between the Indus River on the west and the range of Aravalli Hills on the east are described as the desert soils. They are Regosols of wind-blown sand and sandy fluiratile deposits. These are coarse-textured and derived from the disintegration of rocks in the subjacent areas and blown in from the coastal region and the Indus Valley.

These soils are composed of sand to a depth extending beyond

50 cms. Their A-horizon is weakly developed or absent, as on shifting sand dunes. The soils show no or only a weak B-horizon. Regosols on young dunes are usually yellowish-brown to very pale-brown in colour and may contain considerable amounts of weatherable minerals.

Some of these soils contain a high percentage of soluble salts in the lower horizons. They have a fairly high pH and varying amounts of $CaCO_3$. Being poor in organic matter, they show a low loss on ignition. The fine-textured soils occurring in the Punjab regions would qualify for being classed as Sierozems.

TERAI SOILS

The soils of the Terai region lying at the foot of the Himalayan range possess certain characteristics of their own, which makes it necessary to denote them separately on the Soil Map. These occur all along the foot hills in the northern parts of Uttar Pradesh, Bihar and West Bengal, and are fairly deep and moderately fertile soils. There is considerable evidence that these soils are deposited as a result of their movement through water erosion of the Himalayan range. Lime in the form of fine nodules occurs in these soils in some areas. The surface soils possess a sandy loam or silty loam texture, with the illuviation of the finer particles into the B-horizon being marked. The lower strata are, however, formed of water-worn, rounded stones and gravel of miscellaneous rocks moved down from the hill ranges. The layers of the rounded stones may extend to depths varying from 100 cms to over 300 cms from the surface and contain different proportions of the soil material. These layers of stones make the soils permeable, but the natural formation of these soils at the foot of the hills, where due to high moisture regime and continual seepage of water from the hill ranges, the lands are subject to waterlogging. The excessive soil moisture and the fertility of the soil have caused excessive growth of rank vegetation and weeds, but once drainage is improved and cultivation adopted, the soils become highly productive.

BROWN HILL SOILS

These soils occur often in hilly regions and under moderately heavy vegetation. The soils are formed over a variety of parent rocks, but in the sub-Himalayan region, where they are characteristically located, these materials are sandstones, grey micaceous sandstones and shales. The original vegetation was mainly coniferous and the soils have developed to their present condition after the removal of the original vegetation. They form a system of low foot-hills corresponding to the Shiwalik ranges and are composed entirely of tertiary, and principally of the upper tertiary formations. The rainfall in the region of their occurrence is moderately high and in

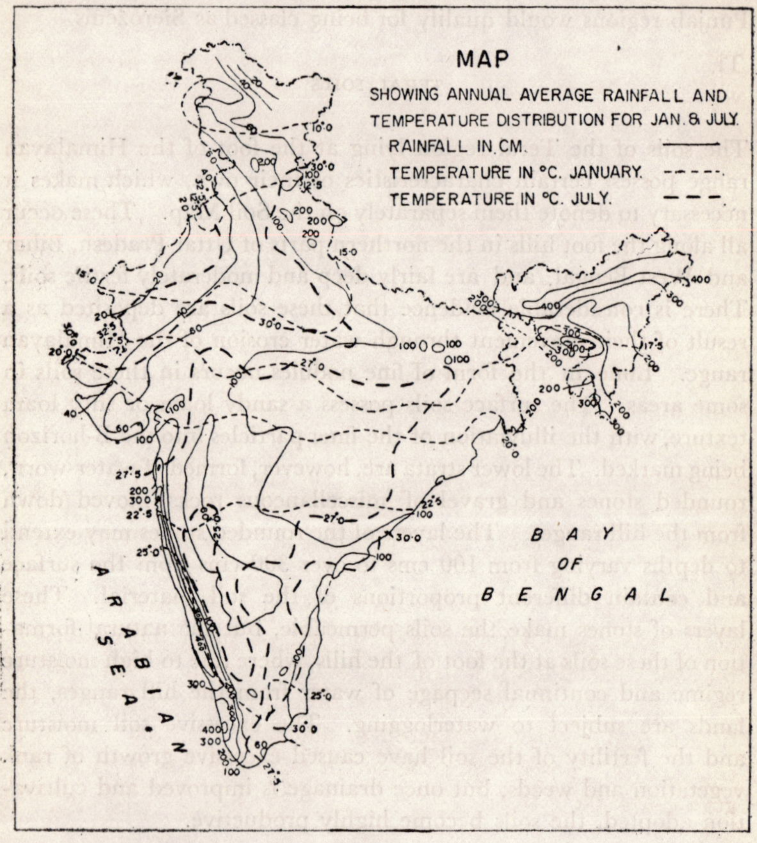

FIG. 22. Rainfall map of India.

the range of 40″ to 70″ per annum. The surface soils are dark-brown in colour and loam to silty-clay loam in texture and moderately rich in organic matter. The B-horizon is fairly deep, 50 to 80 cms, with texture in the silty loam to clay range and compact. These horizons are lacking in free lime and have a pH around neutrality slightly on the acid side. The lower horizons below 100 cms and extending to 200 cms and over are slightly acid (pH 6.0 to 6.5) and composed of compact grey and dark-brown clay loam. Below this layer are the weathered parent material of sandstones and mixed sandstones and shales.

SUBMONTANE SOILS

These are soils found in the sub-Himalayan region under forest vegetation of the coniferous type. The natural vegetation may consist of deodar (Cedrus deodar), spruce (Pica morinuda), blue pine (Pinus excelsa) and chir (Pinus longifolia). The rock types are of the usual varieties of hard and soft sandstones and shales, characteristically occurring in the upper Himalayan regions. The rainfall is usually high, being in the region of 70″ to 100″ per annum, and the accumulation of organic matter in the surface layers, and the lack of free lime could qualify these soils to be grouped under the brown podzol soils. The top surface layer which may extend from 10 to 15 cms is of dark-brown and black sandy loam with loose, undecomposed organic matter. The next layer extending to a depth of 50 cms is a dark-brown sandy clay loam rich in humus. The pH of both these layers is in the region of 5.0. The succeeding layer, to a depth of 15 cms, is of a lighter colour representing eluviation and, below this the layer going to a depth of about 100 cms is brown to reddish-brown, and compact sandy clay. The lower layers are mixed with the weathered gravel of the parent material of sandstones and hard shales. The pH of all the horizons is on the acid side and free lime is absent. Analysis of these soils indicates that the top layers are siliceous in nature, that there is an illuviation of sesquioxides at a depth of about 60 cms, that calcium is conspicuously absent from the profile, and that the C/N ratio is good and balanced.

MOUNTAIN MEADOW SOILS

High up in the Himalayan region at elevations about where conifers or other trees do not grow, and also on the slopes with a southern aspect at lower elevations exist soils of moderately shallow depth, developed from the sandstones and shales, forming the rock mass. The soils because of their relatively thin formation and the poor profile development may also be considered as skeletal soils of the Himalayas. The vegetation is mainly grass, and the growth of the grass helps to afford the binding material to prevent the loss of soil through wind and snow action. The build-up of organic matter from the grass roots is assisted by the relatively low temperatures prevailing in the region. The soils in the nature of their formation vary a great deal in their texture and structure, being admixed with varying proportions of the partly weathered gravel, rock pieces of sandstone, and shaly material.

SALINE AND ALKALI SOILS

Saline soils represent the group of soils that are characterised by the occurrence of a high proportion of soluble salts, usually, the chlorides and sulphates of the alkali bases. These can occur in a variety of soils, and are found among the groups of red, black and alluvial soils. No differentiation between these different groups is attempted in the map, and the saline soils demarcated in it reflect the other general characters of the soils predominantly occurring in the vicinity. In a number of cases, as in the case of soils occurring close to the sea, the salinity is due to the action of sea water and the salts therein. In other cases, the enhanced salinity is due to the removal of salts from soils in higher topographic situations and their accumulation in low-lying areas in an extensive manner. This kind of feature is fairly common in undulating land over which red soils occur. In the case of the areas with black soils, the soluble salts usually occur in the strata below the surface, and the introduction of irrigation—or other causes whereby the water-table is raised—causes the general rise of salts to the surface levels. In the areas with alluvial soils, as in the Gangetic plain, the lack of drainage causes a general rise of the soluble salts, which occur as a thick inflorescence on the surface layers in the dry months. The electrical conductivity of these soils is usually 4 mmhos, and over their pH in the region of 8.0 to 8.50.

The soils affected by alkali have a high pH which may range

between 9.0 and 10.5, and the exchangeable sodium may be over 15% of the total exchange capacity. In these cases the physical condition of the soils is also affected considerably due to dispersion of the clay colloids, which affects their cultivation properties and their cropping capacity.

PEATY AND PEATY SALINE SOILS

The peaty saline soils are developed from brackish water sediments and contain a good amount of sulphides, principally composed of pyrites (FeS_2). These may also be classed as very acid alluvial soils and they are found to occur characteristically in the back-water areas of Kerala State in the south-western tip of the Indian peninsula, and are located at the junction of the hill streams flowing from the western coastal range of hills and the backwaters of the sea. When not drained, their pH may be only slightly acid. By oxidation and hydrolysis of the sulphur compounds contained in the soil, sulphuric acid is formed in varying amounts and the pH goes down to somewhere between 3 and 4, and in extreme cases, a low pH of 2.0 is not unusual. This may occur in some layer or else throughout the profile. Another important characteristic of these soils is their high content of free aluminium and iron, which are formed by the breakdown of clay through the action of sulphuric acid. Accumulation of organic matter in layers of peaty matter results in the lack of oxidation and poor drainage conditions.

Small areas of peat accumulation at the surface occur in the northern parts of Bihar, and they are characterised by the presence of a high amount of organic matter, low base saturation, and a low pH.

SKELETAL SOILS

The skeletal soils are so designated because of the poor thickness of the soil and the lack of characteristics of profile development in them. Such soils usually occur over sandstones of the Vindhyan formation. The soils are coloured pale-brown to dark-brown and may have a thickness varying from 7 to 15 cms, except in hollows between the rocks and in positions sheltered from erosion where they may be thicker. The soils are easily disturbed through erosion and, in consequence, the development of any horizonation is hardly noticeable. The texture of the soils is usually light, being sandy loam to loam, and reflects the origin of the soils from highly siliceous rocks. In other areas, where granite and other igneous rocks have given rise to skeletal soils, the reason for the formation of the thin soils is attributed to the resistance to weathering of the minerals in the rocks and the balance between disturbance through erosion and soil formation is in favour of the existence of shallow soils without any marked profile development. The poor nutrient supply from such soils sustains only a hardy type of natural vegetation, and the soils have low value for cultivated crops.

The skeletal soils in the Himalayan and Ladakh regions are composed of thin soils overlying fragmented sandstone and shale material. The low rainfall, the poor vegetation, the lack of cover, and the existence of conditions unfavourable to intensive weathering have militated against development of deep soils. The factors of wind erosion have also left their imprint by the removal of even the thin soil formed on them, so that in a great measure the rock material remains exposed on the surface.

REFERENCES

Sahasrabudde,, D.L., and Narayana, N., *Agricultural Geology of India and Physical Properties of Soils*, Deccan Book Stall, Poona, 1947.

Raychaudhuri, S.P., *Final Report of All India Soil Survey Scheme*, I.C.A.R., New Delhi, *Bull. No.* 73, 1957.

Raychaudhuri, Agarwal, Datta Biswas, Gupta and Thomas, *Soils of India* I.C.A.R., New Delhi. 1963.

CHAPTER VI

SOIL SURVEYS AND SOIL CLASSIFICATION

In any carefully planned agricultural economy, it is extremely important that the land is developed according to its potentialities so as to derive the maximum and lasting advantage from such land. An assessment of the potentialities of land requires a study of the soil and its associated characteristics through scientific methods, involving soil survey, and interpretation of the data thus obtained. Such a survey will yield fundamental information on the nature of the soil, its texture, depth, structure of different horizons, and erosion susceptibility, besides giving information on drainage and related conditions. Based on this information, sound farm plans, and soil management practices to be followed, can be drawn up. Such information can prevent soil loss through bad land use and erosion, wrong cropping measures, development of saline-alkali conditions and similar undesirable results, which can ultimately cause financial loss to the farmer.

A practical soil survey of any area, whether small or big, involves scientific study of the soil through various stages by well-trained staff with the necessary field equipment and backed by adequate laboratory facilities.

A practical soil survey will include studies on (i) land features and physical relief; (ii) physical characteristics of soil, slope and degree of erosion; and (iii) drainage, salinity, alkalinity, and erosion hazards.

The information can be used in classifying soil into different groups based on their characteristics and the drawing up of soil maps indicating the location, distribution and extent of such distribution of each group of soils recognised and described. In India, many States have organisations for carrying out soil surveys to meet the requirements of different programmes and projects, but the pattern of their organisation or programme of work is extremely variable. On an all-India basis, the Government of India have organised the All India Soil Survey and Land Use Scheme, and this organisation has been functioning since 1956. This central organisation has four regional centres located at Delhi, Calcutta, Nagpur and Bangalore, each of these centres representing the four major soil groups in the country, viz. Alluvial soil, Red soil, Laterite soil, and Black soil. The centre at Delhi deals

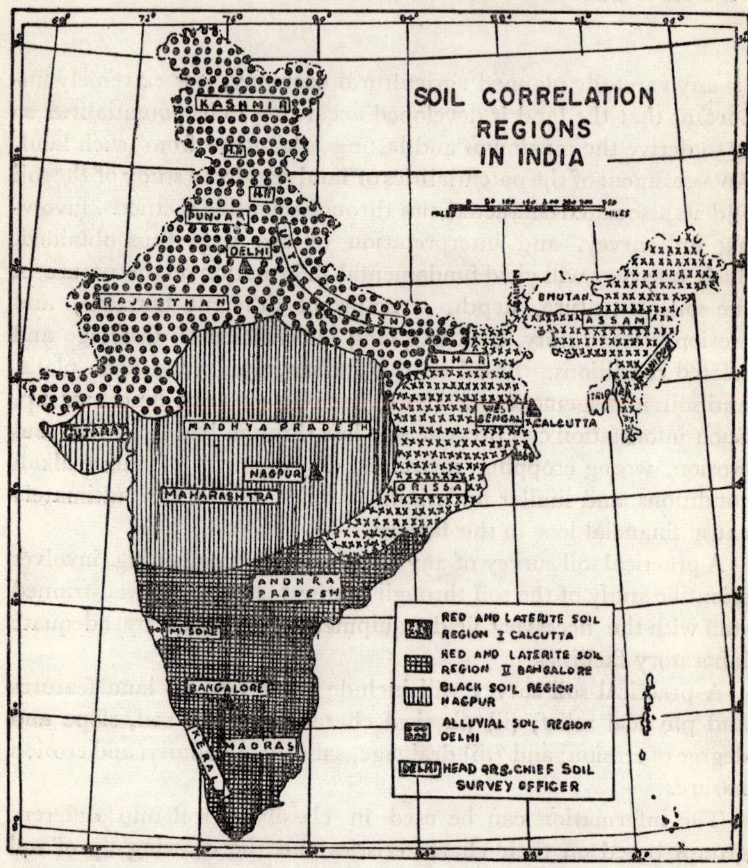

FIG. 23. A map of India showing the four Soil Correlation Regions and their centres.

largely with survey of the soils of the northern region including Punjab, Uttar Pradesh, Delhi, Rajasthan, as also the soils of Himachal Pradesh, and Jammu and Kashmir. The Nagpur regional centre deals with the soils of the central region including Madhya Pradesh, Maharashtra and Gujarat. The eastern soils region is covered by the centre located at Calcutta and operates in Assam, West Bengal, Bihar and Orissa, while the Bangalore centre covers the southern region comprising Andhra Pradesh, Mysore, Madras and Kerala.

SOIL SURVEYS

Soil is the natural medium for the growth of land plants. There are many forms and kinds of soil. The characteristics of soil in any particular place result from the combined influence of a number of factors. These factors include climate, the action of living matter upon the parent rock material, influence of topography and relief, and the period of time these forces have been acting to form the soil. The effect of the cultural environment and man's use of soil are also some forces which have an impact upon the soil development.

Soils must be separated out into individual kinds to study their characteristics and predict their potentialities for use. Classification of soils is an important method of organising the available knowledge and information about soils and the facility of their being easily recognised or remembered by workers. Such a classification of soil has to draw upon experience with the use of the soils, and also the results of study and research. Classification of soils into recognisable groups helps us to see the relationship among soils and to formulate principles for judging their production capacity.

An individual unit of the soils that are distributed on the land surface, can be considered to be a three dimensional body of the landscape that supports plants. Its upper surface is the surface of the land, its lower surface is marked by the lower limits of the soil-forming processes and its sides are boundaries with other kinds of soil. This unit has, therefore, a combination of both internal and external characteristics with certain ranges of properties and characteristics.

Many kinds of soil occur in our country. The characteristics of each kind can be learned through observation and investigation. The history and potentiality of a soil are contained in these characteristics, which individually and collectively influence the behaviour of the soil towards cultivation, crop production and allied operations. Soils are both landscapes and profiles. Hence a study of soil profiles is very important to study not only their inherent characters, but also their extent or distribution on the land surface. This, then, is the purpose of soil survey, i.e. to study

FIG. 24. Soil Survey field work requires recording the information pertaining to soil, the land features, erosional aspects and also vegetation conditions.

the soils, characterise them and determine and depict their distribution in the form of soil maps.

Soil surveys cover a number of activities. They include mapping, classification, interpretation, field and laboratory characterisation for soil mapping and correlation, map compilation and publication. Soil surveys furnish soil maps and interpretation data, and also information needed in research and educational programmes.

The purpose of surveys is to determine the nature and location of each kind of soil. Soil surveys are classed as (1) Reconnaissance surveys; (2) Detailed surveys, or (3) Detailed reconnaissance surveys. Each one of these differs in the method, the details, scales and the resulting precision in mapping.

Reconnaissance Surveys

In reconnaissance surveys, the scale of mapping is usually small and in India it is usually in the scale of 1″ = 1 mile (or the 1 : 50,000 in metric scale), the scale in which topographical maps of the Survey of India for a great part of the country are available. The information mapped in such surveys relates to the kind of soils occurring in the area as defined by soil series. The boundaries between the mapping units, which in this case are the series or an association of series, and in some cases, phases of the series also, are plotted from observations of the soils made at definite intervals in the course of the traverse.

Detailed Surveys

In the case of detailed surveys, the scale of base maps used is larger and may range from 4″ = 1 mile to 16″ = 1 mile. The scales of the maps available in some parts of the country for this purpose are even larger. The base maps used for these surveys are village cadastral maps, which give details about individual fields and indicate their boundaries. In view of the large scale of the maps used, soil information of a detailed manner can be recorded. The details and information about soils mapped include, besides the soil series, the texture, depth, slope percentage and erosion class. All these particulars are depicted in the form of symbols or abbreviations in the form of a fraction which is the "mapping unit." The symbols within the boundaries of a mapping unit indicate the characteristics of the soil within that boundary and also the other items of information about the texture of the top soil, its depth, and information about the terrain. The boundary of the mapping unit is determined by detailed traverse of the field taking auger borings at definite or stated intervals and recording the details thus observed on the base map.

Detailed Reconnaissance Surveys

A detailed-reconnaissance survey involves a combination of the two types of survey over an area, where each type of survey is carried out over different parts of the area surveyed depending upon the intensity of such survey required by each part. Thus in

FIG. 25. Study of Soil profiles requires detailed observations, among others, on the depth of each soil horizon, organic matter in the horizons, texture, structure, permeability of different horizons, root-penetration, nature of parent material and parent rock.

an area which is composed partly of agricultural land and partly of forest lands, the agricultural land may be covered by detailed survey, and the forest land by reconnaissance survey.

SOIL SURVEY METHOD

Soil surveying, as already stated, consists of the examination, classification and mapping of soils in the field. The soils are studied by examining the profiles in pits, usually of size 3 ft by 3 ft and 5 to 6 feet in depth or down to the hard layer of parent rock or water table, whichever is met with earlier. These profile pits are located at suitable distances apart, depending upon the nature of the survey. In detailed surveys, they are located at short intervals of, say, 1 to 2 miles apart and in reconnaissance surveys farther apart. The soil profile, as discussed in Chapter IV, is made up of different soil horizons which are distinctly recognisable layers, and each of these layers is studied carefully to determine the characteristics that affect plant growth, like colour, texture, structure, permeability, root penetration, presence of salts, rock material, etc.

Soils with similar profile characteristics, other than texture of the surface horizon, are grouped together and designated as a soil "series." A soil series, therefore, includes all soil types having about the same kind, thickness and arrangement of layers in the soil profile. The same soil series may have different types depending upon the texture of the surface layer. A soil series takes on the name of the place or location where it is first found and identified or studied. To cite one example, Ooty is the name of a soil series first recognised near Ooty town in Nilgiri District. The series consists of deep friable to moderately friable, well-drained soils which have a pH on the acid side with a dark reddish brown surface layer of usually loamy or silty loam texture with a crumb structure, followed by a reddish-brown or dark-brown subsoil. A characteristic stoneline of 10″ to 12″ depth of weathered gneissic rock material occurs about 18″ from the surface. A typical profile of an Ooty series soil is illustrated in Fig. 26.

The name "Ooty series" has been applied to soil with profile characteristics according to the above description. The same series name is given to all soils having similar profile characters or varying within a narrow range, wherever else they are found.

Variations in the texture of the surface soil of a particular soil series are distinguished and indicated by "types" of this series. Thus, a series may have different types depending upon the texture of the top soil, the other profile characters remaining the same.

The grouping together of the types, based on slope of the terrain, and the erosion features which are the "phases," are the mapping

Profiles of some Established Soil Series

Ooty Series

FIG. 26. Picture of the profile of a well-recognised soil, showing the separation of the different horizons and the description of each horizon. Such specific descriptions help to fix soil series, which can be recognised by soil workers in any part of the country. Fixing the properties of soils and describing soil series helps to recognise the potentialities of each kind of soil for agricultural and other uses.

units. The following examples serve to indicate the series, the variants of the series into types, followed by the mapping units.

TABLE XIII

Series	*Types*	*Mapping units*
Ooty	Ooty silty loam	Ooty silty loam, deep, rolling phase. Ooty silty loam, deep, eroded undulating phase
	Ooty loam	Ooty loam undulating, deep, eroded rolling phase
		Ooty loam, shallow, eroded undulating phase
		Ooty loam, shallow, eroded hilly phase

The mapping of the soil series types and phases as indicated by "mapping units" is done in detailed surveys only. In reconnaissance surveys, as already indicated, only the soil series are mapped. However, when two or more kinds of soil are so mixed that they cannot be shown separately on a map on the scale used, they are mapped together. The areas comprising such soil series are called "soil associations" or "soil complexes" depending upon the difficulties in delineating the component soil series mapped in the area and indicating their boundaries separately. In some cases, prominent and distinguishing characteristics like stoniness or gravelliness, occurring over sufficiently large areas, may be delineated as "phases" in reconnaissance survey maps.

In soil mapping, in detailed surveys, as followed in our country, the mapping units are indicated by symbols which are suitable abbreviations to indicate the series, texture, depth, prevailing slope and erosion characteristics of the soils of the area. The following standard abbreviations and symbols are used for denoting each of these factors.

SOIL SERIES

The soil series names are abbreviated into three-letter units which are adequately indicative of the series named and these abbreviations are used in the soil mapping unit. Thus "Oot" is the abbreviation used to indicate the Ooty series in the concerned mapping units.

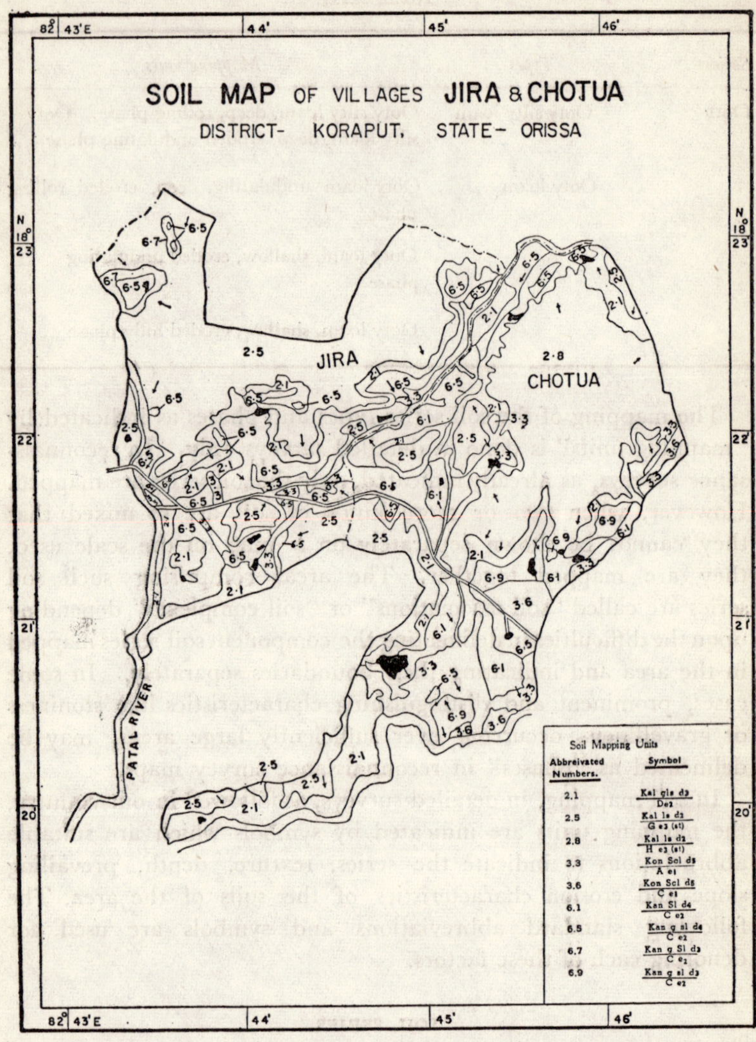

FIG. 27. A typical detailed soil map depicting the occurrence of different soils, their types and other details which help to recognise the character of the soil and erosional features. Such details facilitate interpretation for land use capability.

SOIL TEXTURE

The following abbreviations indicate the different textures most commonly met with:

s—sand	fs—fine sand
sl—sandy loam	l—loam
si—silty	sil—silty loam
c—clay	sicl—silty clay loam
sic—silty clay	cl—clay loam

SOIL DEPTH

The soil depth and the relative symbols used for each depth class are as described in Table XIV.

TABLE XIV

Depth range	*Depth class symbol*	*Description*
Above 36″	d_5	Very deep
18″ to 36″	d_4	Deep
9″ to 18″	d_3	Moderately deep
3″ to 9″	d_2	Shallow
3″ or less	d_1	Very shallow

The soil depth is taken to represent the depth of the solum which in some cases may be only the A-horizon and in many others the A plus B horizons.

SLOPE CLASSES

The slope of the land is normally measured by the hand level (Abney's level). The reading obtained by this instrument is the slope in degrees of inclination (θ), and the corresponding value for the tangent of this angle ($\tan \theta$) is expressed as a percentage or the difference in elevation in feet for each 100 feet horizontally. Thus a land slope of 45° is one of 100% since the difference in elevation of two points, 100 feet apart, horizontally, is 100 feet.

The slope gradients have significance in indicating the susceptibility of soils to erosion and for guidance in formulating recommendations for land use and management. Soil slope classes are drawn up, giving certain ranges of gradients within each class. The following nine slope classes are found to be useful in our country and applicable to the wide range of soil conditions occurring here. The symbols which are used in the soil mapping units and their descriptions, to indicate the respective slope classes, are given against each slope range.

TABLE XV

Slope range	*Slope class symbol*	*Description*
0—1%	A	Nearly level
1—3%	B	Very gently sloping
3—5%	C	Gently sloping
5—10%	D	Moderately sloping
10—15%	E	Strongly sloping
15—25%	F	Moderately steep to steep
25—33%	G	Steep
33—50%	H	Very steep
Over 50%	I	Excessively steep

EROSION CHARACTERISTICS

Features of erosion are largely judged by the conditions existing in the field, the comparison being made with a typical uneroded soil profile. The following are the four classes of erosion found convenient to indicate on the soil map.

TABLE XVI

Erosion class symbol	*Description*
e_1	No erosion or slight erosion
e_2	Moderate erosion
e_3	Severe erosion
e_4	Very severe erosion

Slight erosion (e_1) may be associated with mild sheet erosion; moderate erosion (e_2) with sheet and rill erosion; severe (e_3) with excessive surface erosion, exposing the subsoil or material of the substrata and recognised by the tendency to form small gullies. Very severe erosion (e_4) may be associated with extensive gullying and cutting up of the land by water erosion.

SOIL MAPPING UNIT

The above symbols and abbreviations combined with the abbreviations to indicate the soil series are used to denote a "mapping unit" which represents the soil characteristics and associated features within the boundary marked on the soil map. Thus, a symbol $\frac{\text{Oot sil } d_4}{\text{C}e_1}$ in a detailed soil map indicates that, within the boundary marked on the map, the soil belongs to the Ooty series, with a silty loam texture in the surface soil which is deep. The area is gently sloping and there is slight or no erosion.

Such a study of a detailed soil map, besides giving a picture of these soils, gives additional information about the texture of the surface soils, their depth, as also the prevailing slope and erosion characteristics within the boundaries of each of the marked areas. Additional information about drainage facilities and permeability of the soil in any particular area is obtained from the known properties of the soil as defined by the soil series. A study of the soil map of an area covered by reconnaissance survey gives a picture of the soils, as defined by the soil series, occurring in the area, their extent and their distribution.

While the reconnaissance soil map of an area gives the user of the map a general picture of the kinds of soil occurring in different parts of the area, which can be useful for assessing the broad potentialities of the soils of the area, the detailed soil map gives considerable additional information about the cropping potentialities of different parts of the area, the hazards of their cultivation, and the likely management measures needed for proper conservation of the soils and their fertility. The important characteristics which help to decide whether the soil and the prevailing topographic conditions would permit of agriculture, or compel their being kept under pasture or permanent vegetation are brought out in adequate detail. The data furnished in these surveys and the associated soil maps, therefore, afford material for detailed knowledge of the soil and for further interpretation in regard to the use of the land.

CHAPTER VII

LAND CLASSIFICATION

SOIL SURVEY INTERPRETATION

THE purpose of soil survey interpretation is to provide people with the best possible information about soils in a form that is directly useful to them. Interpretation of soil survey data is a very important part of using the information for practical agriculture. Besides this, the information has other uses like road making, airways construction, industry, urban area development, sewage disposal, recreation, etc. In fact, the information made available gives the basic knowledge about the soils of an area, which, with additional specialised information, can help in interpretation for the various other uses indicated above, besides those of a purely crop-producing nature. In particular, with regard to the use of the data to agriculture, the primary object is to help the interpretation of the data for land capability classification. Land use or land capability classification is an interpretative grouping of soils based mainly on the information in respect of (1) the inherent soil characteristics, (2) the external land features, and (3) the environmental factors that limit or tend to limit the use of land. Information on the first two items is provided by standard detailed soil surveys. The soil units established after field and laboratory studies and correlation are the ultimate management units which afford specific information about ability of the soil to respond to plant growth, use and management.

INFLUENCE OF SOIL FACTORS

As indicated in the earlier sections, classificational soil units provide information on the colour, texture, depth, structure, consistence, permeability, soil reaction, root distribution along the depth of the profile and nature of the parent material. Besides this, detailed laboratory analysis can give information in regard to the types of clay minerals present in the soil, which gives informa-

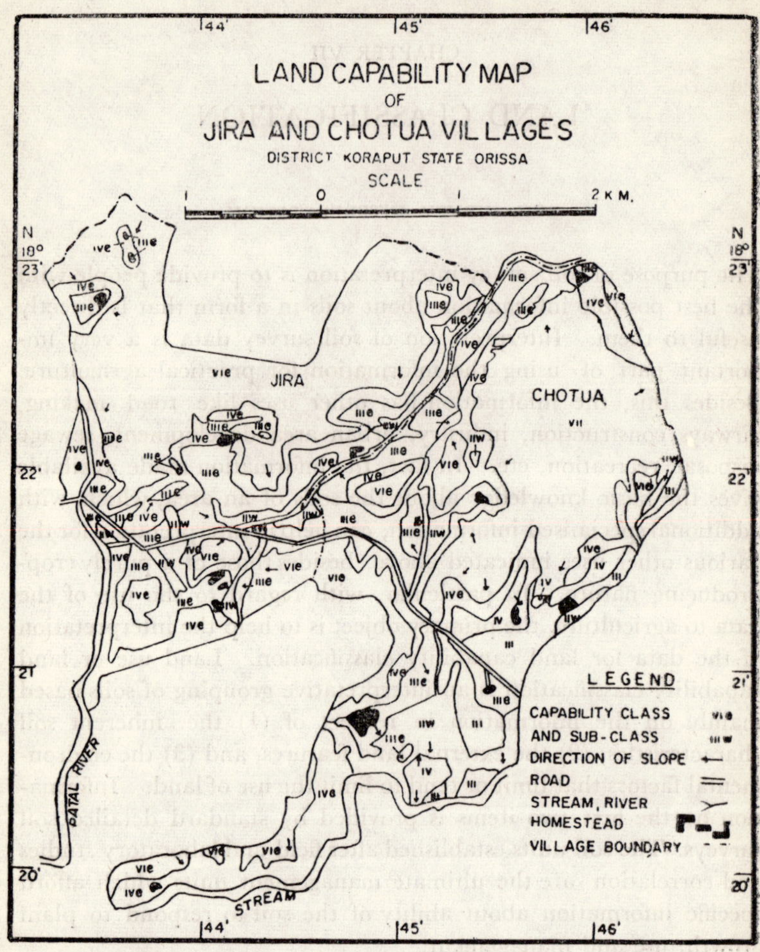

FIG. 28. A Land Use Capability map of the area whose soils have been surveyed and mapped in detail. Interpretation of the soil details given in the soil map (see Fig. 27) has enabled preparation of this land use classification map. It is often convenient to combine a detailed soil map and a land use capability map of the same area in one and the same map as this enables a clear picture to be obtained of both the soil properties and the land use class. Present land use can also be indicated by appropriate symbols on such maps, as this would enable comparison being made of the pattern of present use and the use that the land should be put to for optimum results.

tion about the performance of the soil under cultivation practices, reserves of certain mineral nutrients and their possible supply.

Abbreviated number	*Symbol*	*Full description of soil mapping unit*
2.1	Kal gls d_3 De_2	Kallupada gravel loamy sand, moderately deep moderately sloping (5-10%) moderate erosion.
2.5	Kal ls d_2 Ge_4	Kallupada loamy sand, shallow, steep (25-33%) severely eroded, stony phase.
2.8	Kal ls d_2 He_3 (St)	Kallupada loamy sand, shallow, very steep (33-50%) severely eroded, stony.
3.3	Kon scl d_5 Ae_1	Kona sandy loam, very deep, nearly level (0-1%) slightly eroded. (terraced).
3.6	Kon scl d_5 Ce_1	Kona sandy clay loam, very deep, gently sloping (3-5%) slightly eroded.
6.1	Kan sl D_4 Ce_2	Kangrapada sandy loam, deep, gently sloping (3-5%) moderately eroded.
6.5	Kan g s_1d_4 Ce_2	Kangrapada gravelly sandy loam, deep, moderately sloping (3-5%) moderately eroded.
6.7	Kan g s_1d_3 Ce_2	Kangrapada gravelly sandy loam moderately sloping (3-5%) moderately eroded.
6.9	Kan g s_1d_3 Ce_2	Kangrapada sandy loam, moderately deep, moderately sloping (3-5%) moderately eroded.

The parent material gives an idea of potential nutrient status of the soil. The soil colour speaks of organic matter content and the state of leaching and of hydration. Texture and structure influence the air-water balance in the root zone and movement of water. Slow permeability of the sub-surface layers resulting from presence of clay pans, hard calcium carbonate ($CaCO_3$) or compact limestone layers, etc. can influence root development and cause continuous or periodic waterlogging, a high water-table or flooding. Root distribution determines the depth of soil material up to layers where root penetration is inhibited. Limitations of soil depth control root development and affect moisture retention which in turn influences run-off and also surface soil loss. The soil reaction determines the state of base saturation as also the balance of available plant nutrients, or excess of some constituents leading to toxicity or deficiency which can affect plant growth and crop yield.

INFLUENCE OF LANDSCAPE FEATURES

Besides the above inherent soil characteristics, landscape features like slope and erosion conditions can limit the safe and productive use of soil. The slope gradient and length of slope profoundly affect the rates of rain-water flow, soil removal and indirectly the amount of water absorbed by the soil. The degree of erosion indicates the decrease of productivity and strongly suggests future use and treatment. Climatic factors can place limitations on land use. One unfavourable feature of sufficient intensity in the soil may so limit its use that extensive treatment to protect the soil would be necessary. Several minor unfavourable features may collectively become a major problem and thus limit the use of the soil. Among these limitations are stoniness, salinity and alkalinity, high water-table, hard clay pan, etc. which unless corrected or removed affect land capability.

INFLUENCES ON SOIL CONDITIONS AND CROP PRODUCTIVITY

Classification of soil units into capability groupings enables one to get a picture of (1) the hazard of the soil to various factors which cause soil damage, deterioration or lowering in fertility, and (2) its potentiality for production. The various factors mentioned above individually and collectively control in differing measures the considerations for placing a soil in a particular group. While it is possible to measure the influence of the various external factors which cause deterioration of the soil or its cropping capacity, it is not so easy to assess readily the inherent soil factors controlling the cropping potential. This latter factor can be judged mainly from information of the yields under different management practices and cropping patterns and rotations. A soil with a capacity to give a high yield, besides a capacity to grow a large variety of crops, will naturally qualify to be grouped in a better class than another soil with lower performance under similar management conditions. However, the information available under Indian conditions is yet inadequate to differentiate the soils into different capability groups based on cropping capacity. As more information in respect of the relationship of soils (soil series) and their cropping capability becomes available, it should become possible to base the groupings more precisely.

LAND CAPABILITY GROUPINGS

According to the commonly accepted practice of land use capability groupings, lands are divided into eight capability groups. Of these eight groups, four represent groups of land fit for agriculture and the remaining four are not suitable for cultivation, but suited for pasture, forestry and wild life. The definitions of these groups, or classes which are numbered serially I to VIII, are as follows:

Group I

Under Group I, comes very good land that can be cultivated safely with ordinary good farming methods. It is nearly level land, has deep, productive, easily-worked soils, and is not subject to more than slight water or wind erosion. It is well drained and is not subject to damaging overflows. It is suited for intensive cropping and responds well to manuring and fertilizer applications. In some places the land for crop use may require the use of fertilizers and lime, cultivation of green manure crops and following of crop rotations.

Group II

Group II is good land that can be cultivated with easily applied special practices. Some of its variations are gentle slopes, moderate susceptibility to erosion, soils of only moderate depth, occasional moderate overflow and moderate wetness easily correctable. Soil conservation practices may include bunding, strip cropping, crop rotations that include grasses or legumes, vegetated water disposal areas, cover or green manure crops, fertilizers and manures.

Group III

Group III consists of moderately good land that can be used regularly for crops in a good rotation, with intensive treatment. Such lands are characterised by moderately steep slope, high susceptibility to erosion, moderate overflow, slow or very slow subsoil permeability, excessive wetness, shallow depth to bed-rock, hard pan or clay pan, sandy, very sandy or gravelly soil with low moisture capacity and low inherent fertility. More intensive measures to protect the soil like bunding or terracing and attention to drainage in case of wetness will be required in lands of this group.

FIG. 29. Wrong use of land leads to rapid deterioration. This is a Class VII land which should not have been disturbed or cultivated but put under permanent vegetation to protect against soil loss. Land use classification is a scientific guide to proper use of land as suited to its capability.

Group IV

Group IV constitutes fairly good land whose cropping use is restricted by natural features such as steep slopes, moderate to severe erosion, unfavourable soil characteristics or adverse climate. It needs intensive treatments, special cropping systems or special practices to protect it against erosion and to conserve moisture. Terracing as a soil conservation measure, and intensive drainage measures in case of severe drainage impedence are some of the measures indicated, besides growth of special cover crops and intensive management practices.

Group V

Land in group V is not suited for cultivation but is suited for perennial vegetation including pasture and some types of forest vegetation. Cultivation is not feasible because of one or more factors, such as wetness, stoniness or some other limitation. The land is nearly level and not subject to more than slight wind or water erosion. Examples are those of the bottom lands subject to frequent overflow that prevents the normal production of cultivated crops, but are well suited to growth of pasture for grazing.

Group VI

Group VI is land subject to moderate limitations under grazing or forestry use. It is too steep, subject to erosion, shallow, wet, dry, or otherwise not suited to cultivation, but well suited to grazing or forestry. Cultivation of plantation crops like tea and coffee, which provide for good ground cover with moderate disturbance of the soil through cultivation, may be permitted under certain conditions.

Group VII

Land in Group VII is subject to severe limitations or severe hazards under either grazing or forestry use. It is not suited for cultivation. It is very steep, eroded, stony, rough, shallow, dry, swampy or otherwise unfavourable, but can be used for grazing or forestry if carefully handled.

Group VIII

Group VIII comprises land of such unfavourable characteristics as to be unsuited for cultivation, grazing or forestry. It is suited for wild life, recreation, or watershed protection uses. It includes such areas as marshes, deserts, deep gullies, high mountain land and very steep, rough, stony barren land.

LAND USE MAPS

Interpretation of the soils data from a soil map enables the preparation of a land use map, where the differing areas which require to be placed under the different land capability groups are indicated. In such a land use map, the different groups are distinguished by colouration according to the following standard pattern.

Group I	———	Light green
Group II	———	Yellow
Group III	———	Red
Group IV	———	Blue
Group V	———	Dark green
Group VI	———	Orange
Group VII	———	Brown
Group VIII	———	Purple

Land Capability Subgroups

The soil capability groupings II, III, and IV for agricultural use could be divided into subgroupings on the basis of the following limitations, and these indicated on the land use maps to denote the presence or occurrence of special limitations or factors of additional hazards. These subgroups are intended to indicate: (1) risk of erosion or post-erosion damage designated by the symbol (e); (ii) wetness, drainage or overflow (w); (iii) root zone limitations (s); and (iv) climatic limitations (c). The concerned suffixes are attached to the group symbols in the maps. Where soils have two kinds of limitation, both the concerned subgroup symbols can be indicated under the group symbol, the dominant one being shown first. If two kinds of limitations are essentially equal, the subclasses have the following priority: e, w, s and c.

Soil Capability Units

To facilitate a better understanding of the potentialities of the soils under survey, soil capability subgroups can be further divided into capability units to indicate soils which respond in a similar

FIG. 30. Both orchards and annually cultivated crops exist on this land, which could be classified under Class II. The soil conditions on this land are such that a variety of annual crops and also perennial crops could be raised on them. Lands classified under a particular class can be put to use to suit the cropping and management practices of all higher classes. But the converse practice of cultivating with crops on lands which should be under permanent vegetation, is not possible.

way and require similar management practices although they may have soil characteristics that put them into different soil series. For example, a soil capability subgroup area may have different parent materials so that the soil management practices, such as application of fertilizers in a cropping schedule, will be different. In such a case, the capability subclass is divided into two units and thereby distinguished.

From the foregoing, it is evident that information in regard to the various soil properties, factors of topography, soil limitations, etc. require to be considered individually and collectively in assessing the capability of lands for use and management. A

TABLE XVII

RELATIONSHIP BETWEEN SOIL FACTORS, LAND FEATURES AND CAPABILITY GROUPING

	Characteristics of land features and soil factors	*Limitation for agricultural and related uses*	*Identification with general type of land*	*Classification under land capability groups and subgroups*
	1	2	3	4
1	Level land with deep, well drained soil of satisfactory texture and structure, free from harmful factors like excessive lime, acidity, alkalinity or salts; fit for a variety of crops; responsive to manuring. The soil texture in these cases may usually range from fine sandy loam to loam or silty loam and well furnished with soil organic matter and plant nutrients. These soils have capacity to maintain physiological availability of water.	Few or no limitations of soil depth, erosion, water table, subsoil barrier, overflow or wetness, etc.	Fertile, high yielding alluvial lands. Lands naturally occurring on level terrain on high river banks and those with similar characters with facilities for irrigation.	Normally Group I land. This class will go up higher with the effect of enviornmental and other limitations.
2	Gently to moderately sloping land with deep soil of fairly satisfactory texture and structure; subject to occasional overflow or wetness easily correctable; slightly affected by salinity and such harmful factors. Moderately susceptible to erosion or slightly to moderately eroded. Sandy loam to loamy soil overlying heavier subsoil, with moderate permeability and drainage. Fine sandy loam to loam soil overlying sand or gravel, deep soil but rapidly permeable. Silty loam to loam on clay or silty clay moderately deep, slowly permeable.	(a) Susceptible to moderate erosion. (b) Slight drainage impedence. (c) Slight marshiness. (d) Droughtiness. (e) Moderate condition of salinity and acidity slightly affecting crops. (f) Nutrient reserve poor.	(1) Rainfed lands for dry and wet crops in red and black soil regions. (2) Medium to heavy textured soils of shallow valleys in moderate rainfall regions. (3) Medium to heavy textured soils under bunds, on gently to moderately sloping terrain. (4) Bottom lands subject to occasional overflow, easily correctable. (5) Foot hill soil quite deep, overlying porous substrata of sand or gravel or boulders.	Group II lands. Erosion hazard will need classification as IIe, root-zone limitation as IIs and drainage difficulties as IIw.

3	Gently to moderately to strongly sloping lands, deep to moderately deep soil of satisfactory texture, subject to frequent overflow or wetness, presence of harmful factors including salinity or alkali, high water-table, etc. Moderately affecting crop growth. Soils of heavy texture subject to severe erosion hazards or having adverse effects of past erosion. Coarse to medium texture soil moderately deep underlain by sand or gravel. Silt loam to loam, poorly drained moderately permeable over very slowly permeable substrata (clay pan, fragipan, etc.)	(a) Very susceptible to erosion. (b) Drainage impedence rather marked. (c) Marshiness. (d) Droughtiness and root zone limitation marked. (e) Salinity and alkali hazards. (f) Nutrient reserve poor. (g) Poor moisture-holding capacity.	(1) Heavy-textured soil of expanding type clay minerals with gentle slope in black or mixed red and black soil regions. (2) Soils with subsoil barrier in the alluvial, red or black soil regions. (3) Lands in alluvial or black soil region with high water-table, affected by salinity or alkali that reduce crop yield. (4) Bottom lands that remain wet or are subject to frequent overflow. (5) Bunded lands on moderate to strong slopes in the red and laterite soils. (6) Terraced land for paddy cultivation in strong slopes in the hilly and mountainous region with moderate rainfall.	Group III land. Erosion hazard or past erosion will need classification as IIIe root zone limitation as IIIs and drainage difficulty as IIIw.
4	Strongly sloping to steep land, deep to shallow soil favourable texture; soils severely affected by salinity or alkali permitting only salt tolerant crops to grow. Soil under severe erosion; sandy or concretionary soil with very low moisture-holding capacity. Highly saline organic soils. Gravelly or concretionary soil overlying gravelly or concretionary subsoil, rapidly permeable, well drained. Sand to loamy sand overlying sand or coarse material, well drained, rapidly permeable.	(a) Severe past erosion. (b) Severe root-zone limitation. (c) Pronounced marshiness. (d) Bad droughtiness. (e) Strong salinity or alkali hazard. (f) Lack of nutrient reserve in the surface. (g) Moisture-holding capacity very poor. (h) Aridity of climate restricting cropping.	(1) Concretionary soil of the nature of *bhata* soils of M.P. or *dabba* soil of A.P. (2) Coastal alluvium (sandy) or sandy deposits on river banks; other sandy stretches as in cultivated fields. (3) Sandy soils of arid and semi-arid regions. (4) Alluvial or black soils of semi-arid regions, severely affected by salinity or alkali. (5) Kari (saline peaty) soils. (6) Terraced paddy lands on steep slopes. (7) Bunded lands with stone revetments on steep slopes.	Group IV land. Erosion needs classification as IVe, root zone limitation as IVs, wetness as IVw.

Table XVII (*continued*)

	1	2	3	4
5	Level or nearly level land with stony or rocky soil; marshy soils, etc.	(a) Cultivating implements cannot be worked. (b) Crop management practices cannot be applied.	Lithosolic soils of central India; marshy lands. Stony or rocky lands. Bottom lands along stream banks, subject to frequent flooding.	Group V lands.
6	Steep to very steep land with deep to shallow soil; shallow to very shallow soils on gentle to steep slopes; soils with severe erosion hazards or very severe past erosion. Soils with extreme salinity and alkali hazards; soils under severe climatic restrictions.	(a) Very rapid run-off attaining high velocity causing deep cutting. (b) Stoniness and erosion condition hindering cultivation of arable crops by application of usual management practices. (c) Low moisture-holding capacity and nutrient reserve. (d) Severe salinity or alkali hazard.	(1) Steeply or very steeply sloping lands of the hilly and mountainous regions with deep to shallow soils. (2) Severely eroded lands of red and black soils or other soil groups. (3) Very shallow soils with concretions of kankar or stones on the surface. (4) Desert sands and sand dunes. (5) Coastal sand dunes. (6) Lands severely affected by salinity or alkali. (7) Lands subject to acute marshiness, including coastal mangroves.	Group VI lands. With increasing severity of the hazards, the lands are to be placed under Group VII
7	Bad lands, rock out crops, etc.	(a) Severe root limitations.	(1) Rocky escarpments (2) Cherty and very stony ground with shallow soil interspersed with sheet rock.	Group VIII

ready guide to indicate the relationship of these various factors and the grouping of the lands into the different groups and important subgroups is furnished in Table XVII. A perusal of the characteristics of soil and land features given here can help easy identification by the field workers of the general types of land associated with each of the eight capability groups described earlier.

CHAPTER VIII

PLANT FOODS IN THE SOIL

ELEMENTS REQUIRED IN PLANT NUTRITION

THE composition of animal and plant matter shows a great deal of similarity. Animals, however, must rely on other animals or plants as a source of food so that their own living substance may continue to exist. Animals are completely dependent, in the final analysis, upon the plant kingdom for life. However, green plants can exist and increase independently of any animal life. All that is required is a supply of water, carbon dioxide, and several mineral elements to make the green plant a completely self-sufficient organism in the presence of light. The raw materials that the plant uses in the manufacture of its tissues then assume a role of no small importance, and a knowledge of the functions of these materials should be of first importance to persons engaged in the care and production of plants.

The composition of plants in general indicates that a large number of elements go into their make-up. Some of these elements are found more commonly and in greater abundance than others. The elements: carbon, hydrogen, oxygen, nitrogen, phosphorus and sulphur are the ones of which proteins, and hence protoplasm, are composed. Besides these six, there are ten elements known to be essential for plants. They are: calcium, magnesium, potassium, iron, manganese, copper, boron, zinc, molybdenum, and chlorine. These elements, along with phosphorus and sulphur, usually constitute what is known as the plant ash, or the material which remains after the burning off of the carbon, hydrogen, oxygen and nitrogen. Each of these sixteen elements plays a specific role in the growth and development of a plant, and when present in insufficient quantity, growth, and hence crop yield, may be seriously reduced.

Because some of these elements are present in large amounts in plant materials and are also taken up in large quantities from the soil by plants, they are described as macro-nutrient elements. Included among these are carbon, hydrogen, oxygen, nitrogen,

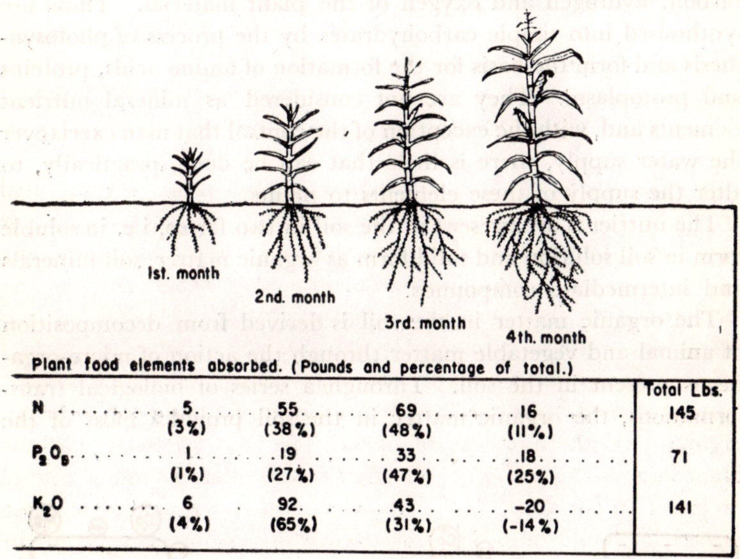

	1st. month	2nd. month	3rd. month	4th. month	Total Lbs.
N	5 (3%)	55 (38%)	69 (48%)	16 (11%)	145
P_2O_5	1 (1%)	19 (27%)	33 (47%)	18 (25%)	71
K_2O	6 (4%)	92 (65%)	43 (31%)	-20 (-14%)	141

FIG. 31. Plants utilise mineral nutrients from the soil and the quantities absorbed vary with the age of the plant.

phosphorus, potassium, sulphur, calcium and magnesium. Of these, nitrogen, phosphorus and potassium are the *major plant nutrients* which are usually supplied to the soil in the form of fertilizers. The elements, e.g. sulphur, calcium and magnesium are referred to as *secondary plant nutrients* and their supply to the soil to supplement the normal supplies is made in the form of amendments and frequently in the form of fertilizers supplying the major plant nutrient elements. Besides these, seven other elements are present in relatively small amounts in plant material, and are also taken up from the soil in small amounts and, accordingly, they are described as *micronutrient* elements or minor elements of plant nutrition. These are iron, manganese, boron, copper, zinc, molybdenum and chlorine. It is necessary for successful crop growth, that all the above sixteen elements are available to the crops in adequate quantities. If any one of these elements is lacking or is present in insufficient amount, the growth suffers, and is restored only when the insufficiency is made up to an adequate extent.

Atmospheric carbon dioxide and soil water are the sources for carbon, hydrogen and oxygen of the plant material. These are synthesised into simple carbohydrates by the process of photosynthesis and form the basis for the formation of amino acids, proteins and protoplasm. They are not considered as mineral nutrient elements and, with the exception of the control that man exerts over the water supply, there is little that can be done, practically, to alter the supply of these elements to plants.

The nutrients are present in the soil in two forms, i.e. in soluble form in soil solution and solid form as organic matter, soil minerals and intermediate compounds.

The organic matter in the soil is derived from decomposition of animal and vegetable matter through the action of micro-organisms present in the soil. Through a series of biological transformations, the organic matter in the soil provides most of the

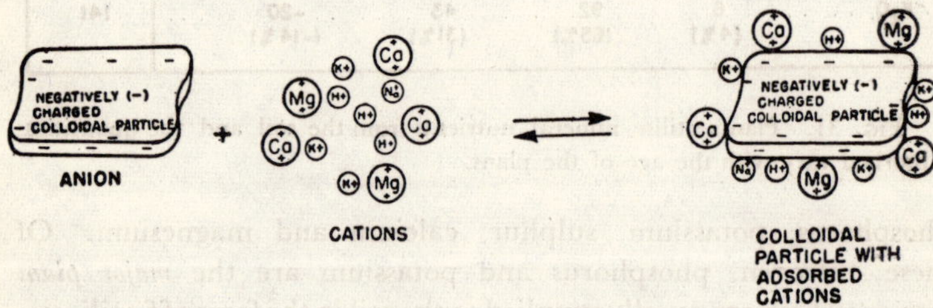

FIG. 32. Colloidal clay in the soil acts as a large anion, that is, it adsorbs cations and holds them in exchangeable forms that are available to growing plants. The negatively charged clay, in the presence of cations in the solution, adsorbs the cations to satisfy the capacity of the negative bonds. In arid regions, the cations will be dominated by sodium and in humid regions, by hydrogen. Ammonium is also capable of being adsorbed on the colloidal clay but the amount is usually small and under certain conditions, the ammonium is oxidized to nitrate. Colloidal humus also acts in a similar manner as colloidal clay.

nitrogen, a part of the phosphorus, and a part of the sulphur to the the growing crops.

The mineral fraction in the soil supplies most of the other nutrients to the crops. These nutrients are contained in minerals form in the sand and silt fractions of the soils. The process of conversion of primary mineral nutrients into forms available to the crops, otherwise known as *mineralisation*, is a very slow one. The clay fraction contains two important components, viz. the clay minerals and the hydrous oxides of iron and aluminium, both being derived from the weathering of rocks.

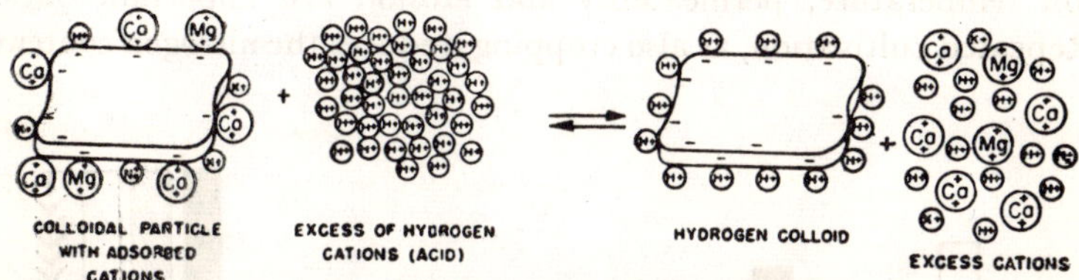

FIG. 33. A diagram illustrating cationic exchange in soils. On the left is a colloidal particle with adsorbed cations of calcium, magnesium, potassium, sodium, and hydrogen. When this colloidal particle is treated with an excess of hydrogen ions (cations) in solution, by mass action the H_2 replaces Ca_2, Mg_2, Na_2, and K_2, leaving a hydrogen-saturated colloid, and the other cations remain in solution. This reaction is also reversible, depending upon the relative number of each cation.

Ammonium (NH_4) is also an exchangeable cation but it does not remain long as NH_4. Under aerobic conditions and with the moisture content between the field capacity and the wilting point, the NH_4 changes rapidly to NCl.

A colloidal humus particle in the soil also reacts in the same manner as colloidal clay.

ROLE OF MAJOR PLANT NUTRIENTS

Nitrogen

Nitrogen is a vitally important nutrient element, the supply of which in the soil may be markedly influenced by man. Nitrogen is present abundantly in the atmospheric air of which 80 per cent is gaseous nitrogen. Gaseous nitrogen is, however, an inert element and it must be chemically combined with other elements in the form of compounds before it can be of use as plant food. The nitrogen in the soil is present chiefly in the organic matter of the soil and to a much smaller extent in inorganic mineral forms. An average soil may contain from 500 to 5000 lbs of total nitrogen in the top soil in an acre of land. Most of this nitrogen is, however, held in tight chemical combination and is released only very slowly through bacterial decomposition of soil organic matter. The combined mineral forms in which nitrogen is assimilated are the nitrate (NO^{-}_{3}) and the ammonium (NH^{+}_{4}) ions.

Organic nitrogen is gradually depleted from the soil through oxidation and the activities of micro-organisms. The rate of this loss depends upon various factors of which soil texture, aeration, soil temperature, permeability and erosion are important ones. Repeated cultivation, as also cropping, reduces the nitrogen content

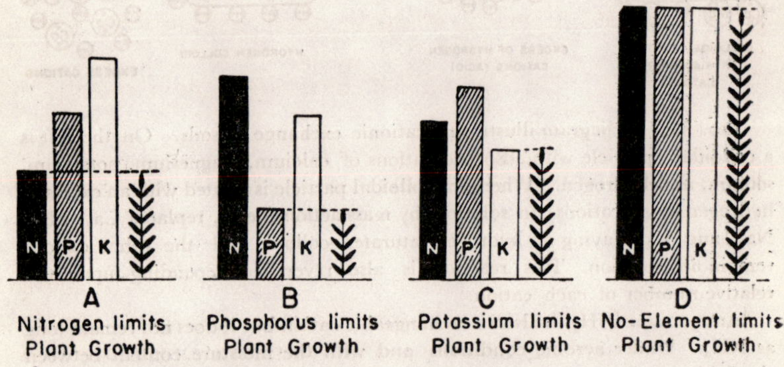

FIG. 34. A diagram illustrating limiting factors in crop production. At "A" nitrogen is in least relative supply and, therefore, limits plant growth. At "B" phosphorus is limiting plant growth, at "C" potassium, and at "D" neither N, P, or K is limiting plant growth.

in soils. It is, therefore, necessary that nitrogen in one form or another should be applied to cultivated fields to replace the losses. By the natural activities of certain nitrogen-fixing organisms present in soils and in the nodules of the roots of legumes, like pulses, sunnhemp, etc., which are able to assimilate the nitrogen of the atmosphere and fix it in the form of organic compounds, the natural loss of nitrogen from the soil is replaced to a certain extent. The development of chemical processes, whereby the nitrogen of the atmosphere is combined with hydrogen and oxygen and converted into compounds which can serve as rich sources of nitrogen, have revolutionized the methods of nitrogen supply to the soil and the replacement of natural losses. These processes have built up the enormous industries in different countries of the world which are serving to supply nitrogen to the soil in the form of concentrated fertilizers.

The important functions that nitrogen plays in crop growth may be enumerated as follows:

1. It is an important constituent of proteins which, in their turn, are elaborated into plant protoplasm.
2. It is an integral part of chlorophyll and hence is associated with the production of green matter in plants.
3. Supply of nitrogen to plants increases yield of leaves, fruit and seed.
4. Adequate supply of nitrogen increases protein content of grains and food crops.

Deficiency in the supply of nitrogen to plants is easily noticed by characteristic symptoms. A stunted yellow appearance is indicative of a nitrogen deficient plant The yellowing or chlorosis, usually appears in the lower leaves, the upper leaves remaining green. In cases of severe nitrogen shortage, the leaves usually "fire" or turn brown, beginning at the leaf tip and progressing along the mid-rib until the entire leaf is dead.

Nitrogen is lost from the soil through oxidation, denitrification and leaching, and this loss naturally leads to lowering of the fertility of the soil. Attempts to raise the nitrogen level in soils suddenly through a heavy application of manures or artificials will lead to wastage. It will be a good policy to keep the nitrogen at a moderate level through regular applications of organic manures supplemented with artificial ones, where increases are needed.

Phosphorus

The phosphorus in the soil is present in two forms: (a) organic and (b) inorganic. The relative proportions of these two forms in soils are highly variable with the organic fraction varying from 20 to 60 per cent of the total phosphorus. The organic forms belong to the three groups: (a) phytin and phytin derivatives, (b) nucleic acids, and (c) phospholipids. Plants utilise phosphorus in the mineral form and hence the organic form has to mineralise before this element is available.

The inorganic forms of phosphorus occur usually as compounds of calcium (fluorapatite) or iron and aluminium phosphates. All these forms are highly insoluble and are rarely of much use to the plants.

Since plants can absorb phosphorus only when it is present in a soluble or weakly held condition, the condition in which the phosphorus is present in the soil is a very important factor. In the presence of free iron, aluminium and calcium, soluble forms of phosphorus tend to be converted or "fixed" into highly insoluble forms and become unavailable. Thus, when soils are highly acidic, iron and aluminium of the soil become soluble and react with the phosphates to form highly insoluble iron and aluminium phosphates. Similarly, under alkaline conditions in the presence of excess of calcium carbonate, highly insoluble compounds of calcium and phosphorus may be formed. Hence both acidic and alkaline reactions in soil restrict and reduce phosphate availability and a pH range between 6.5 and 7.5 is ideal for phosphate availability.

A somewhat similar situation occurs in the use of organic phosphorus too. The availability of organic phosphorus to crops is best between pH 6.0 and 8.0, and will be limited beyond these pH ranges.

THE ROLE OF PHOSPHORUS IN CROP GROWTH

The importance of phosphorus in plant and animal nutrition is well recognised. It is present in seeds in larger amounts than in any other part of the plants, although it is found extensively in the young growing plants. Like nitrogen, it is a constituent of every living cell. Sufficient quantities of phosphorus are necessary for normal transportation of carbohydrates in the plants, like conversion of starches into sugars. Also, they help in the assimilation of fats in plants. The presence of phosphorus aids in the uptake of potassium and counteracts the effects of excess of nitrogen.

The most obvious effect of phosphorus is on the root system of plants. Phosphorus-starved plants have stunted systems, which reduce their feeding zone and thus the plants will be unable to withstand adverse conditions. The stimulation of root development with additional amounts of available phosphorus is especially valuable in heavy and fine-textured soils where root development is naturally restricted, and also furnishes the plant with the means

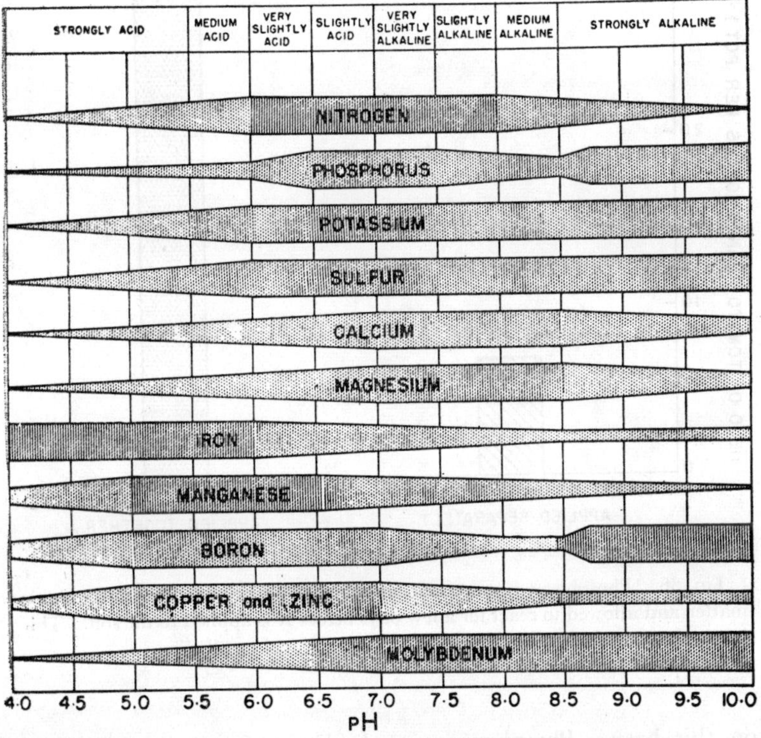

FIG. 35. The relationship between soil pH and relative availability of essential elements from the soil. The wider the bar, the more the relative availability.

of withstanding drought conditions. In conditions of drought, it will be easily seen that crops well supplied with phosphorus appear green and growing, when crops not furnished with this nutrient show signs of wilting.

Phosphorus hastens the ripening process. So, seed formation begins soon and the crops may mature several days sooner than when the phosphorus is deficient. Phosphorus is essential for seed formation, and its effect on maturity of crops may be explained

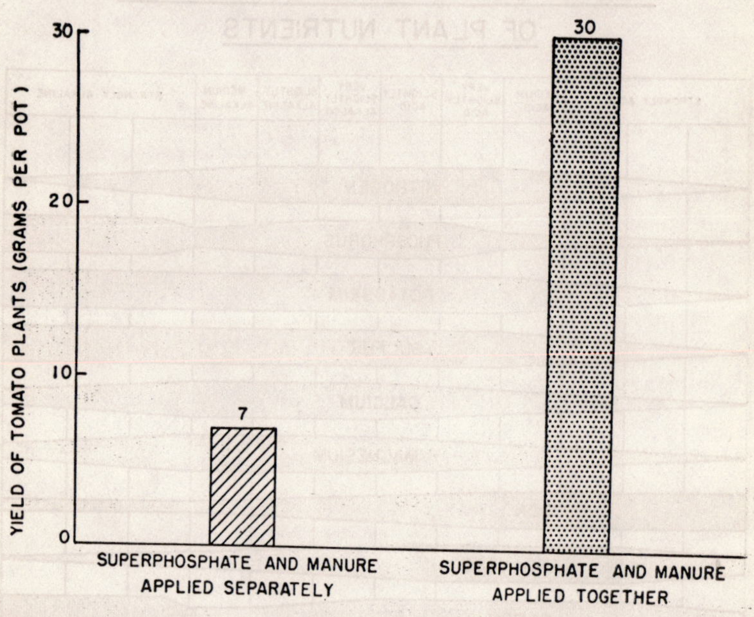

Fig. 36. Phosphorus is more efficient when it is first added to fresh organic matter and allowed to react for a few days before it is applied to the soil. This is especially true when the soil is either strongly acid or alkaline.

on this basis. Phosphorus-starved plants may mature late and their quality may be lowered, as phosphorus builds up resistance in plants, and in this respect offsets the disadvantages of excess of nitrogen which stimulates vegetative growth and renders the plants susceptible to diseases. Phosphorus deficiency causes stunted growth and restricted tillering of cereals. The tips of cereal leaves develop purple patches as in the case of maize, sorghum and ragi. The stalks remain slender. The yield of grain and fruit is considerably reduced.

Potassium

Potassium occurs in both soluble and insoluble forms in the mineral components of soils. A large proportion of soil potassium is in the primary minerals, like the micas and feldspar which occur in the sand and silt-fractions. A small part of the soil potassium is in a readily available condition. This constitutes the soluble potassium in the soil solution which forms 1 to 5 per cent of the total and the exchangeable potassium present in the exchange complex of the clay colloids, which forms 1 to 3 per cent of the total. There is always an equilibrium between the soluble and the exchangeable potassium. Thus, when plants remove some soluble potassium from the soil solution, more potassium moves out from the exchangeable form to become soluble.

A part of the potassium gets "fixed" in the clay minerals like illite, montmorillonite and vermiculite. Such a reaction takes place readily on drying. Such "fixed" potassium will become available again only gradually. The supply of potassium from the soil is also affected by the level of other nutrients such as calcium, nitrogen and phosphorus.

Soil potassium is lost through cropping, erosion and leaching. The loss through cropping varies according to the crop and may range from 25 to 150 pounds per acre. Since only a small part of the land in India receives extra applications of potassium through fertilizers, much of the loss through cropping is made up by fresh releases from the mineral fractions of the soil. In some soils, this release may not be enough and so applications of potash fertilizers may be necessary. The types of soil which require such applications are organic soils, acid soils and highly leached soils.

In plants, potassium is needed most for intense growth, and at flower and fruit initiation and at setting.

ROLE OF POTASSIUM IN PLANT GROWTH

Potassium performs many functions in the physiological process in plant life. One of the more important functions usually attributed to this element is its effect on the plant synthesis of carbohydrates and proteins. Potassium is essential for the production of starch, sugar and other carbohydrates, and the translocation of starch and other materials within the plant. It aids in the reduction of nitrates in the plant preparatory to protein synthesis and is also believed to be essential for the development of chlorophyll and in the synthesis of oils and fats and albuminoids.

The general vigour and tone of the plant which permits the plant to have greater resistance to disease is due to the presence of potassium. Crops which do not receive sufficient potassium are more susceptible to disease and this effect is noticeable when there is excess of vegetative growth due to excess of nitrogen.

Potassium increases plumpness in grain, thereby increasing the quantity of the product, and makes the stalk and straw of plants

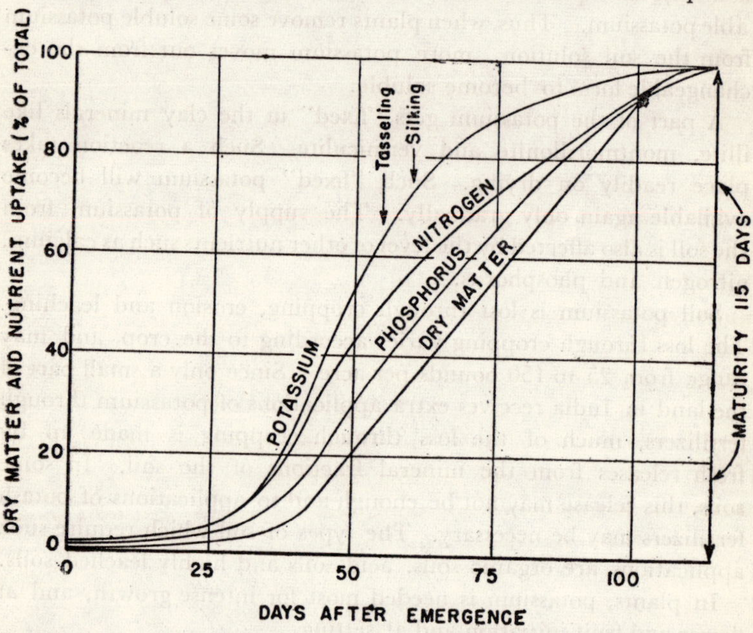

Fig. 37. The uptake of nutrients by maize plant and the increase in dry matter in relation to the number of days after emergence. Note that the increase in N+P+K requirements is highest at the time of tasselling and silking.

more rigid, thus preventing lodging to a certain extent. Where a deficiency exists, the addition of potassium improves the quality of tobacco, coffee and potato. It also facilitates the growth of the sugarcane crop.

Potassium deficiency becomes obvious at a rather late stage in plant growth, when it is too late for corrective treatment. The symptoms are yellowing of tips and edges of the leaves and, later, this widens and the edges start drying up. These symptoms appear first in the older leaves and then in the younger ones. The yellowing may also start as yellow spots or dots (as in clover, lucerne etc.) and as streaks (as in maize, jowar, etc). In cotton, mottled yellowish margins on leaves with unopened and partly open bolls are characteristic symptoms.

SECONDARY PLANT NUTRIENTS

Calcium

Calcium is one of the essential elements of plant nutrition, but is grouped among the secondary plant nutrients. It is absorbed by plants as the ion Ca^{++} just as potassium and this may take place either from the soil solution or by the process of contact exchange.

Calcium is present in soils (excluding that added as lime or other fertilizer materials) and owes its origin to the rocks and minerals from which the soil is formed. It is contained in a number of minerals like dolomite, limestone, calcite, apatite and calcium feldspar.

Calcium is an important element in plant growth. It promotes root formation and growth. Its deficiency is characterised by malformation and disintegration of the terminal portion of the plant. It helps to remove toxic compounds in the plant cells. Like potassium, it improves straw and grain formation, and also gives resistance to the plants towards disease.

Calcium tends to be deficient in acid soils and in highly leached laterites and red soils. When liming materials are used to correct soil acidity, calcium deficiency is also made up at the same time.

Magnesium

This element is a constituent of chlorophyll and hence, as with other nutrient elements, a deficiency of magnesium results in characteristic discolouration of the leaves. Premature defoliation sometimes results from magnesium deficiency.

Magnesium is present in soils in primary mineral forms and in other minerals like dolomite, serpentine, etc. The presence of magnesium in plants regulates the uptake of other plant nutrients.

In soils a satisfactory calcium to magnesium ratio is desirable to permit the normal uptake of the elements. Excess of magnesium

Fig. 38. Manganese deficiency in cotton: *Left:* Normal leaf of cotton; *Middle:* Leaf of cotton showing mild manganese deficiency; *Right:* Leaf of cotton with acute manganese deficiency.

or too low a concentration in relation to calcium has undesirable effects on plants.

Among the functions of magnesium in the plant system are those attributed to the formation of oils and fats, and the translocation of starch. Magnesium takes an active part in enzyme systems.

Highly leached soils, which have not been limed, tend to be deficient in magnesium. In such cases, the application of dolomitic limestone is recommended.

Sulphur

Sulphur is an essential element for plants and is present in both organic and inorganic forms in soils. Sulphur is a component of living matter and so, when this material is returned to the soil and converted into humus, a large proportion remains in organic combination. Sulphur tends to be deficient in some areas and symptoms of sulphur deficiency in cotton and tobacco in sandy soils have been reported. Certain plants like crucifers and lilies require large amounts of sulphur, and they respond to sulphur applications in the form of sulphates. Sulphur is also reported to have given increased yields of clover.

Sulphur is found to improve root growth in most plants and helps maintain the dark-green colour of the leaves. Sulphur or sulphates promote nodule formation in legumes and encourage more green leaf and seed formation. Sulphur is applied to the soil indirectly in many forms as fertilizers, ammonium sulphate, superphosphate, potassium sulphate and as gypsum. Sulphur is present in industrial chimney smokes from coal burning factories, and is brought down along with rain water in the sulphate form.

Sulphur deficiencies are apt to develop in areas having low organic matter in soils and where high nitrogenous, sulphur free fertilizers are used. Symptoms are stunted growth, and a pale-green to yellow foliage. In some plants, say, turnips, redness develops in the lower leaves which later spreads all over.

MICRO-PLANT NUTRIENTS

The micro-plant nutrients are also referred to as trace elements or as minor elements, in consideration of the trace amounts in which they are required by plants. These elements are iron, manganese, zinc, copper, boron, chlorine and molybdenum.

The micro-nutrients though needed in very small quantities are essential for plant growth. In their absence, growth may be deficient or even stopped altogether. The plants develop characteristic deficiency symptoms specific to each deficient element and the recognition of these symptoms enables the deficiencies to be noticed. However, the quantities necessary for normal growth of the plants are small and such amounts are usually present in the soils or added to the soil through impurities in the applied fertilizers. Therefore, regular annual applications may not be necessary. However, corrective doses in the form of soil applications or sprays of salts of the concerned elements would need to be applied if deficiency symptoms are noticed.

TABLE XVIII

Micro-nutrient elements	*Role in plant nutrition*	*Deficiency symptoms*	*Method of correction*
1	2	3	4
Iron	Constituent of respiratory enzymes like cytochrome, catalase, dipeptidase, etc., and thus controls respiration, photosynthesis and other vital activities. Necessary for the formation of chlorophyll in leaves.	Young leaves remain small and become pale-yellow in colour. Deficiencies are particularly noticed in trees and bushes rather than on annuals.	Spraying the foliage with organic iron complex salts, chelates such as Fe EDTA shows improvement. The chelates may also be applied to the soil.
Manganese	Necessary for the formation of chlorophyll, reduction of nitrates, respiration, protein synthesis and formation of vitamin C.	Manganese deficiency results in chlorosis of leaves. The inter-veinal areas are bleached but veins remain green.	Application of 40-50 lbs of manganese sulphate per acre to the soil will correct manganese deficiency.
Zinc	Necessary for certain enzymatic reactions in cells and for the formation of chlorophyll and auxin.	Dwarfed growth and little leaf. Formation of rosettes in beans, citrus, cowpea and apple.	Application of zinc sulphate powder of 4 to 5 lbs per acre or use of 0.5-1.5 per cent spray of zinc sulphate or zinc chelates.
Copper	Takes part in certain oxidation and reduction of enzymatic reactions	Copper deficiency becomes manifest under continued droughts. Chlorosis of leaves, falling off of leaves and twigs are the usual symptoms.	Application of 1 to 5 lbs of copper sulphate powder per acre or use of 0.5 to 1 per cent solution spray.
Boron	Necessary for active cell division, germination of pollen, formation of flowers, fruits and roots and cell formation. Seems to be necessary for the transport of carbohydrates and proteins.	Deficiency is first noticed in growing tips and young tissues where curling and malformation will result.	Application of 5 to 10 lbs borax per acre on average soils. Excess will produce toxicity.

Table XVIII (*Continued*)

1	2	3	4
Chlorine	Not yet established.	Leaf wilt, chlorosis and necrosis.	Application of 1 to 5 lbs of chlorine (ammonium chloride or potassium chloride).
Molybdenum	Necessary for nitrogen fixing micro-organisms and for nitrogen transformation in plants.	Whiptail disease of cauliflower, and other cruciferous plants in which the tissues between the veins die. Blue chaff disease of oats in Tasmania.	Usually applied along with phosphatic fertilizers in small doses. 1 to 3 ounces of molybdic acid per acre.

Table XVIII gives the fundamental details with regard to each of the micro-nutrients.

UPTAKE OF PLANT NUTRIENTS BY DIFFERENT CROPS

An analysis of varieties of crops has been made to determine the amounts of major nutrients taken up by different crops. Table XIX gives the figures in respect of crops grown in this country. These data furnish a guide with regard to the normal uptake of nutrients by the crops and indicate the extent to which these should be made available in the soil to furnish the minimum requirements of the crops.

TABLE XIX

QUANTITY OF PLANT NUTRIENTS REMOVED FROM SOIL BY DIFFERENT CROPS

Crop	*Yield in lbs/acre*	*Quantity of nutrients removed in lbs/acre*		
		N	P_2O_5	K_2O
Rice	2,000	30	20	60
Wheat	1,400	50	21	60
Jowar	1,600	50	13	130
Bajra	1,000	32	20	59
Maize	1,800	32	18	35
Barley	1,000	37	18	31
Sugarcane	60,000	80	15	180
Groundnut	1,700	70	20	40
Mustard	600	20	10	25
Linseed	900	17	11	29
Cotton	400	27	15	40
Jute	1,400	60	30	60
Tea	800	40	12	25
Coffee	800	30	10	30
Tobacco	1,300	84	51	81

Source: Handbook of Agriculture, I.C.A.R., 1961.

METHODS OF DETERMINING NUTRIENT REQUIREMENTS OF CROPS

For a satisfactory growth of crops, the plants must be able to obtain the minimum quantities of major nutrients and, besides these, adequate quantities of the 13 other essential elements needed for normal growth and development. If any one or more of these are in short supply, growth becomes retarded, which reflects in crop yield. It is possible to detect the deficiencies of these essential elements in the soil by several methods. These are briefly stated as follows:

1. Examining the characteristic deficiency symptoms in the leaves and the growing parts of the plants.
2. Testing the soil for available nutrients and determining which of them is below the normally required level.
3. Conducting field trials with different types of fertilizers and salts supplying the secondary and the micro-nutrient elements at different levels and periods of growth and determining the uptake of the nutrients by the entire crop and also by the responses as measured by actual yields.
4. Testing the fresh plant tissues in the field itself to determine the deficient elements.

Each of the above methods, however, has its own advantages and disadvantages. Deficiency of the major plant nutrient elements to meet the needs of the growing crop appears frequently as characteristic symptoms in different parts of the plants and these can be visually noted. While it is easy to make observations of the deficiency symptoms on the growing plants, it is rather difficult to estimate them correctly, as the symptoms for different elements overlap and appear similar. Thus, symptoms of nitrogen, sulphur, manganese and iron deficiencies may appear to be similar. Sometimes, the deficiency symptoms appear too late to be corrected and this is frequently the case with potassium deficiency.

The soil tests give a picture of the level of available major plant nutrients in the soil, as also the occurrence of any adverse characters in the soil such as salinity, alkalinity, etc. These tests can be carried out in centrally located laboratories, and, in the following chapter, details of these have been given. The soil tests are, however, to be interpreted by an experienced person before the results can be made applicable in the field.

One of the most reliable methods of detecting deficiency is by actual field trials, using different levels and combinations of nutrients. This procedure is not only laborious, but also slow for general adoption and as such, can be carried out only on a restricted scale and by careful planning.

Of the different methods indicated above, the last one, viz. that of testing plant tissues in the field, offers potentialities of using the

Fig. 39. *Right*: The series of rice plots, from foreground to background, received varying amounts of N in all combinations, with varying amounts of K *but none of the plots received any phosphorus fertiliser*. Note the complete failure of the rice crop. *Left*: The entire series of plots, from foreground to background, all received 20 pounds per acre of phosphorus fertiliser, but varying amounts of N and K. Note the vigour of rice in all plots.

plant itself as an indicator for assessing the adequacy of the nutrients in the soil for its satisfactory growth. For obtaining such information, analysis of the growing plant or, more easily, the examination of the growing parts of the plant appears to be a reasonable method. To meet this requirement, various plant-tissue testing methods have been developed for both annual and perennial crops and these are described below.

TISSUE TESTS

Chemical analyses of fresh extracts of plant tissues can provide useful information on the nutrient deficiencies occurring under actual conditions of growth in the field. Such tests carried out rapidly and in the field itself have been found very convenient and quick for taking prompt remedial measures. They have been widely used in recent years in the case of maize and beet in the U.S.A., and for sugar-cane and pineapple in Hawaii for assessing nutrient needs of growing crops. Similar tests have also been occasionally used in India in case of perennial plantation crops like coffee, tea and citrus.

Among the widely used techniques for tissue testing are those developed by Thornton[1] and others (Purdue University, Indiana), Morgan[2] (Conn. Agric. Exptl. Station), J.B. Hester,[3] and at Urbana Laboratories, Illinois. The main advantages in the use of tissue testing techniques are:

1. The nutrient deficiency can be detected even before the symptoms appear on the plant so that remedial measures can be taken before it is too late.
2. Indications of the necessity for further soil tests can be obtained in case such information is not already available.
3. The adequacy of fertiliser applications and a check on the methods of application to ascertain if they are effective or need modifications, can be made by such tests on the growing crops.
4. The localised soil nutrient deficiencies in the field and in isolated plants in perennial crops can be detected.
5. The most appropriate time for application of fertilizers for each crop can also be effectively determined.

[1]Thornton, S.F., Connor, S.D., and Fraser, R.R. (1939), "The use of rapid chemical tests on soils and plants as aids in determining fertilizer needs," *Purdue University Agr. Expt. Sta. Cir.*, No. 204.

[2]Morgan, M.F. (1939), "Soil Testing Methods," *The Universal Soil Testing System., Circ Conn. Agric. Expt. Sta. 127:* pp. 1-16.

[3]Hester, J.B. (1941), "Soil and Plant Tests as Aids in Soil Fertility Programmes," *Com. Fert. Yearbook*, 1941, pp. 31-9.

Methods of Tissue Tests

Tissue tests are based on the estimation of the nutritional content of tissue sap extracted from related portions of the plant. A petiole or a leaf blade is selected according to age, and small bits of uniform area are cut out by means of a suitable instrument such as a hand punch. The material is crushed to extract sap or extracted with a suitable fluid like dilute acetic acid, and the extracted sap is tested for the major nutrients N (nitrates), P and K by rapid comparison of developed colours.

Usefulness and Limitations of Tissue Tests

The accuracy of tissue tests is dependent, to a large extent, on uniformity of procedure in regard to sampling, as the composition of sap is directly linked with the age of the tissue, the time of sampling and the growth period of the crop, besides the time elapsed since the previous fertiliser application. Hence, the following points have to be borne in mind in adopting a useful tissue testing procedure:

1. The N, P, K, content of plant sap tends to be higher in plants of tender age than in mature ones.
2. The plant sap tends to have a higher concentration of nutrients immediately after application of fertilizers and hence sampling at such a time will not truly represent the actual needs of the plant.
3. The plant selected for sampling should as far as possible be tender and in vigorous growth and of the same age, such as the third leaf from the apex, or fifth leaf from the apex, and so on.
4. Nutrient deficiencies become manifest mostly at definite stages of the life cycle of the plant such as tillering (for cereals), flowering or seed formation (for perennials). This would be the right time for sampling.
5. In order to get a truly representative sample, a large number of sub-samples from a large population is necessary. Odd plants and plants near borders should be avoided.

The following are some of the disadvantages in the adoption of tissue testing on an extensive scale.

1. A single operator can do only a limited number of tests in the field.

2. The interpretation of the tests requires considerable background experience of the crop and the agronomical practices.

3. The tests are subject to personal errors and absolute estimations are often difficult.

4. The tests cannot be conducted in bad weather.

Deficiency Symptoms

Besides the tests carried out on the plant tissues to examine the deficiencies and inadequacies of the plant nutrient elements, other visual method exist, particularly based on judging characteristic symptoms developed by the plant due to such deficiencies.[1] Deficiencies of major and in many cases, the minor nutrient elements in the plants are exhibited in the development of characteristic symptoms associated with the growth and colour of the growing parts, particularly the leaves and the formation of necrotic spots on them. A careful study of these symptoms, while discriminating between conflicting factors, can enable one to form a fairly reliable judgement based on the existence of and the intensity of such deficiencies in the growing plants. The deficiency symptoms, where developed, are due to physiological inadequacies of some of the nutrient elements and, as similar symptoms may be caused by disease or pathological causes, there is need for a careful discrimination between these two factors. Some of the common symptoms developed in plants due to deficiencies of major nutrient elements are given below in the form of a general key. This key will be found useful for general reference purposes only. However, it has to be borne in mind that each crop develops its own characteristic deficiency symptoms which may overlap and, therefore, when a doubt arises, further reference to the literature is necessary.

[1](1) *Hunger Signs in Crops: A Symposium*, American Society of Agronomy and National Fertilizer Association, Washington D. C., 1949, Pp. xv+390; (2) Wallace, T. (1944), *The Diagnosis of Mineral Deficiencies in Plants: A Colour Atlas and Guide Supplement*, His Majesty's Stationery Office, London, Pp. 48, (3) Chilean Nitrate Education Bureau Inc. (1941), *If They Could Speak*, New York, Pp. 54.

TABLE XX

GENERAL KEY TO NUTRIENT DEFICIENCIES IN PLANTS

I. *Effects, general for the entire plant*	*Deficiency*
(a) Effects generally distributed over the entire foliage:	
(i) Plants pale-green—Lower leaves affected first. Turn pale-yellow and later brown. Stunted growth.	Nitrogen
(ii) Plants dark-green and stunted. Stems and leaves purple or orange-brown.	Phosphorus
(b) Effects localised in parts of foliage: Chlorosis, mottling, necrosis, etc.	
(i) Interveinal portions and tips of leaves become yellow, while veins and lower portions remain green.	Magnesium
(ii) Stunted growth. Margins of leaves having mottled condition, yellowing at the edges to form bands and ultimately drying.	Potassium

II. *Effects confined to younger or current year's growth*	*Defficiency*
(i) Stunted growth. Leaves reduced in size. Growing tips and nodes shortened giving a rosette appearance. Leaves yellow between the veins.	Zinc
(ii) Leaves near the growing tip yellow and small. Internodes shortened. Growing tips affected with dead tissue.	Boron
(iii) Leaves usually normal in size. The mid-rib and veins dark-green in colour, but the interveinal portion pale-green or yellow.	Manganese
(iv) Veins remain green. Rest of the leaf becomes pale green or yellow. Dead tissues may develop at the margins.	Iron

REFERENCES

BEAR, F. E., *et al.*, *Hunger Signs in Crops,* The American Society of Agronomy, Washington, 1949.

SCARSETH, G.D., "Soil and Plant Tissue Tests as Aids in determining Fertilizer Needs," *Better Crops with Plant Food, 25,* 9-17, 1941.

MORGAN, N.D. and WICKSTORM G. A., "Give your Plants a Blood Test," *Better Crops with Plant Food,* Vol. XI, 5, 1956.

LAL, K.N. and RAO, M.S.S., *Micro Element Nutrition of Plants,* Banaras Hindu University Press, 1954.

WALLACE, T., *Mineral deficiencies in Plants,* Chemical Publishing Co., New York, 1953.

HALL, A.D. and SMITH, A.M., *Fertilizers and Manures,* John Murray, London, 1955.

CHAPTER IX

FERTILIZERS AND THEIR USE

A FERTILIZER IS A MATERIAL either of organic or inorganic origin which is applied to the field in order to increase its power of nutrient supply to the crops. Sometimes, the materials of an organic nature which may be obtained from natural sources are referred to as manures, and only the inorganic materials which are frequently of a concentrated nature and prepared by artificial means are referred to as fertilizers. There are other types of materials which are also applied to the field to improve its productive capacity. But these, like lime, gypsum, sulphur, tank silt, etc. do not strictly come under the category of fertilizers and they are better termed as Soil Amendments. The soil amendments may sometimes supply plant nutrients in addition to other functions in improving the soil conditions, as in the case of tank silt, but the term "fertilizers" is used with reference to the materials which supply plant nutrients in a concentrated form to the soil.

Fertilizers do not consist of the plant nutrient elements nitrogen, phosphorus and potassium as such, but they are combined with other elements to form either organic or inorganic compounds. Fertilizer materials are classed as nitrogenous, phosphatic or potassic depending upon the principle constituents in them, although some materials like ammonium phosphate, potassium nitrate, etc. can be placed in more than one of these classes. It is customary to refer to fertilizers and fertilizer mixtures as containing nitrogen (N), phosphoric acid (P_2O_5) and potash (K_2O) instead of Nitrogen (N) Phosphorus (P) and Potassium (K). This method of reference is based upon the long existing practice of expressing the nutrient contents in terms of their oxides, except in the case of nitrogen, and there is a trend at present to express all of them on an elemental basis.

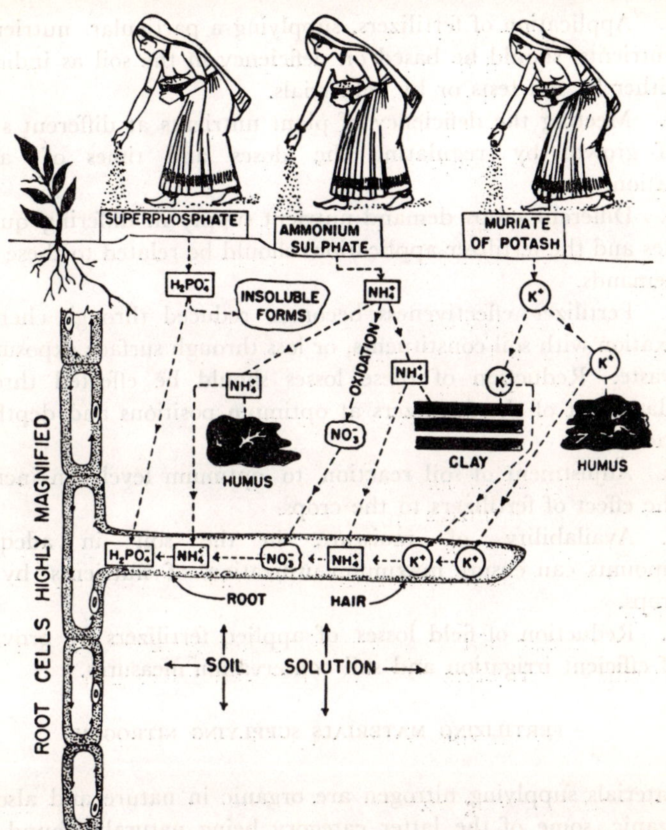

Fig. 40. A simplified diagram showing how a plant obtains phosphorus from superphosphate, nitrogen from ammonium sulphate, and potassium from muriate of potash.

PRINCIPLES TO BE FOLLOWED IN FERTILIZER USE

The best returns from the use of fertilizers can be expected only from their judicious use and by following certain principles with regard to the optimum method of use as also the time of their application to crops. The nutrient supply of the fertilizers is mainly meant for the crop to which they are applied and in most cases the value of the applied nutrients for succeeding crops is considerably reduced, if not completely lost. The following principles can serve as a guide to be followed in the efficient use of fertilizers.

1. Application of fertilizers, supplying a particular nutrient or nutrients, should be based on deficiency in the soil as indicated either by soil tests or by field trials.
2. Meeting the deficiency of plant nutrients at different stages of growth by regulating the doses and times of application.
3. Different crops demand nutrient supply in differing quantities and the fertilizer applications should be related to these crop demands.
4. Fertilizer effectiveness becomes reduced through chemical fixation with soil constituents, or loss through surface exposure or waste. Reduction of these losses should be effected through placement of the fertilizers at optimum positions and depths in the soil.
5. Adjustment of soil reaction to optimum level can increase the effect of fertilizers to the crops.
6. Availability of moisture in the soil in adequate amounts can ensure maximum utilization of nutrients by the crops.
7. Reduction of field losses of applied fertilizers by provision of efficient irrigation and soil conservation measures.

FERTILIZING MATERIALS SUPPLYING NITROGEN

Materials supplying nitrogen are organic in nature and also inorganic, some of the latter category being naturally found and others formed by artificial methods. The atmosphere is the source for all nitrogen. While in the free state nitrogen is an inert material, in organic combination in plant and in animal materials it is useful, Its functions are many and useful. Artificially, it has been possible to combine atmospheric nitrogen with various other elements to give products which are useful as fertilizer materials. A convenient classification of these different materials is based on consideration of the origin and the nature of the materials as follows:

i. Non-synthetic organic materials,
ii. Non-synthetic inorganic materials, and
iii. Synthetic nitrogenous materials.

ORGANIC NITROGENOUS MATERIALS

A great variety of plant and animal by-products of relatively high nitrogen content are available for use as fertilizer materials. They consist of organic waste materials from the farms as also urban localities, like bird guanos, fish scraps, fishmeal, slaughter-house waste, waste from cotton, wool and leather industries, as also the residues obtained after extraction of oil from seeds. Many of these like cotton, wool and leather wastes contain nitrogen in low amounts and also take long periods to decompose in the soil, while others like urban and farm wastes are rarely applied to the soil as such, but are allowed to rot and decompose outside the field before application. Animal residues like dried blood, bird guanos, fishmeal and oil cakes, which have a relatively high nutrient content, are applied to the crops directly.

TABLE XXI

NUTRIENT CONTENT OF DIFFERENT NITROGENOUS MATERIALS OF NATURAL ORIGIN

	N%	P_2O_5%	K_2O%
Farmyard manure	0.5-1.5	0.5-0.8	0.5-1.9
Compost	1.0-2.0	1.0	1.0-1.5
Green manures	0.5-0.8	0.1-0.2	1.0
Groundnut oilcake	7.0-8.0	1.5-1.6	1.3-1.4
Castor cake	5.5-5.8	1.8-1.9	1.0
Mahua cake	2.5-2.6	0.8-0.9	1.8-1.9
Niger cake	4.5-4.7	1.8	1.3
Karanj cake	3.9-4.0	0.9-1.0	1.3-1.4
Neem	5.2-5.3	1.0-1.1	1.4-1.5
Safflower cake (undecorticated)	4.8-4.9	1.4-1.5	1.2-1.3
Sesamum or til cake	6.0-6.3	2.0-2.1	1.2-1.3
Dried blood	10-12	1.0-1.5	0.6-0.8
Fish manure	4-10	3-9	0.3-1.5
Bonemeal-raw	3-4	20-25	—
Steamed bonemeal	1-2.0	25-30	—
Night soil	1.2-1.5	0.8-1.0	0.4-0.5
Ground rice hulls	0.5	0.2-0.3	0.1-0.8

Table XXI gives the details of the nutrient content of the different nitrogenous materials of natural origin and it is seen that while many of them are rich in nitrogen, some contain appreciable amounts of phosphoric acid and potash also.

In the case of organic materials like farm wastes and town wastes, which are obtained in large amounts in urban localities, the materials are subject to a process of "composting" whereby they are converted into a form fit for application directly to the land as a manure. The principle involved is mainly to bring about microbial decomposition of cellulose-rich (high in C/N ratio, and in the region of 80 to 120) materials like dry farm wastes and town refuses by

Fig. 41. Layering night soil (starter) over dry town refuse in Trenches. A scientific method of preparing Urban compost of high quality from town wastes.

combining with starters like dung and cattle urine in the case of farm wastes, and night soil or sewage in the case of town wastes to furnish an end product, the farm or town compost as the case may be, which has a moderately low C/N value—in the region of 9 to 12. The process of composting as in the Bangalore Method, consists in stacking in alternate layers, in a longitudinal manner, the dry refuse with the respective starters in long trenches, so that approximately equal amounts of these are added. Maintaining the moisture conditions at optimum levels facilitates the microbial decomposition. Within a period of 4 to 6 months, the materials thus filled in the trenches become converted into manure of good quality which is fit for application to the soil as a bulk manure.

The organic manures and composts thus prepared contain, principally, nitrogen, besides phosphoric acid, potash and many of the micro-nutrients. These organic manures are valued not only for their nutrient value but also for their other beneficial effects on the land and soil, such as the improvement of microbial activity, providing for better soil aeration, increasing the water-holding capacity and improving the soil structure. They have

Fig. 42. A whole trench filled up with farm compost.

a regulating effect on soil temperature and help to reduce the fixation of phosphate in the soil. Organic materials with a high C/N ratio should not be used in the soil before decomposition or the application of additional amounts of nitrogen to the soil along with them, since they can create deficiency of nitrogen in the soil temporarily and thus affect crop yields adversely.

NEED FOR CARE IN HANDLING FARM MANURES

Considerable losses take place in the collection of cattle manure, principally in the cattle sheds which result in loss of useful fertilizer constituents. Cattle urine is rich in nitrogen and other minerals of value to crop production, and their incomplete collection and storing reduces the value of the manures prepared from farm wastes. The loss of urine occurs mostly from failure to use ample bedding, leakage through stable floors, seepage into the earth floor or drainage from manure heaps.

On the basis of the total plant food in manure, about 50 per cent of the value is in the urine, and hence the quantities largely lost represent the portion which is in the most readily available forms. Improved methods of collecting more completely the cattle urine, when the cattle are housed in sheds, can considerably enhance the value of the farm manures.

Losses occur by volatilization when manure heaps are allowed to lie exposed to the action of sun, rain and wind. The principal losses are in nitrogen and organic matter. Large quantities of ammonia are produced in manure from urea and other nitrogenous compounds and, in the earlier stages of manure decomposition, the ammonia is combined largely with carbonic acid as ammonium carbonate and bicarbonate. These ammonium compounds are unstable and gaseous ammonia is readily liberated, the tendency for this loss being greater with increase in temperature. Protection of manure heaps, whether they are farm composts or urban composts, by keeping them packed in the trenches where they are prepared or by covering them with an adequately thick layer of earth (½ to 1 inch) when they are stored in overground heaps, protects them against the losses indicated above.

PRESERVATIVE TO ENHANCE THE VALUE OF MANURES

Chemical preservatives can be added to manure to decrease nitrogen losses. Their action may be due either to their prevention of decomposition of urea and other nitrogenous compounds, or else the conversion of the volatile nitrogenous compounds into more stable salts. Strong acids like sulphuric and hydrochloric acids are effective preservatives. These chemicals make the manure acid and prevent the biological decomposition of urea and also combine with ammonia to form non-volatile salts. Although these acids are effective preservatives, their cost and the difficulties in handling them limit their practical use.

However, the use of calcium salts of the strong acids like calcium sulphate, calcium phosphate, calcium chloride and calcium nitrate, offers possibilities for conversion of the ammonia present in manures into stable salts. While the use of these salts in a pure form is ruled out, commercial grades of some of these materials can be used with

Fig. 43. Preparation of farm manure composts by the method of sectional trench filling. The first section is completed at one end of the Trench. Partial exposure of the manure provides for the requisite aeration used to start decomposition. The aeration gets progressively diminished as another section gets filled up.

advantage. In particular, ordinary grades of superphosphate (18 to 20 per cent P_2O_5) contain 40 to 50 per cent of gypsum and this is very effective in preventing the loss of ammonia. Application of half to one hundredweight of superphosphate to a ton of farm manure at the time of composting enriches the manure in phosphate besides ensuring a minimisation in the loss of nitrogen. The use of powdered rock phosphate which is less expensive than superphosphate has also been advocated as a means of improving

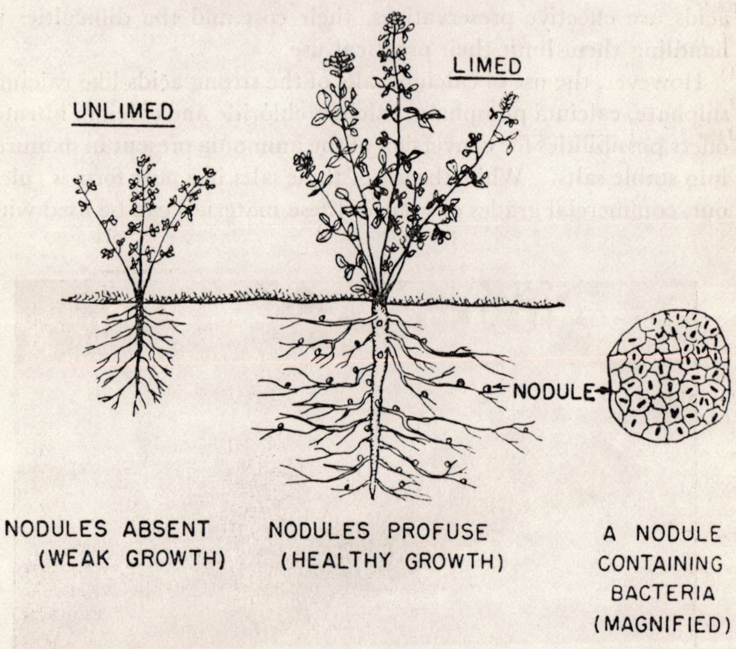

Fig. 44. Lime supplies calcium, sweetens acid soil helps nodulation, and encourages vigorous nitrogen fixation. Every nodule contains lakhs of nitrogen fixing bacteria.

the quality of manure besides enhancing the availability of the phosphate which is in a largely unavailable form in the rock phosphate.

GREEN MANURES

Green manure crops are those which are turned under the soil in a green condition, for improving the soil. There are two types used for this purpose : (i) green leaf manures, and (ii) green manure crops. Manures of the former category are green matter collected from leguminous and other plants grown on plantations or from forests and turned under the soil in cultivated fields. Green leaf and loppings from Karanjor Honge (Pongamia glabra). Thangdi, Avaram (Cassia auriculata), Glyricidia (Glyricidia macculata) and similar plants come under this category. These materials may be applied to the soil in a green condition, or after storing for some days as is done in practice when transported from distances. They serve to improve the soil conditions and, in part, may help to furnish additional supplies of nitrogen.

TABLE XXII

AVERAGE NUTRIENT CONTENT OF GREEN MANURES (FRESH BASIS)

	Percentage content		
	N	P_2O_5	K_2O
Sunn hemp, Sann (Crotalaria juncea)	0.75	0.15	0.50
Black Gram, Urd (Phaseolus mungo)	0.85	0.18	0.53
Green gram, Mung (Phaseolus aureus)	0.72	0.18	0.52
Horsegram, Kulthi (Dolichos biflorus)	0.33	—	—
Clusterbeans, Guar (Cyamopsis tetragonoloba)	0.34	—	—
Dhaincha (Sesbania aculeata)	0.62	—	—
Cowpea, lobia (Vigna catiang)	0.71	0.15	0.58

Green manure crops are usually leguminous, grown on the land and turned under at a suitable time. These plants, if properly inoculated and grown under favourable conditions, build up in their plant matter nitrogen collected from the air. These crops, when incorporated into the soil, add nitrogen to the soil's previous supply. Sunn hemp (Crotalaria juncea) and others of the Crotalaria species, dhaincha (Sesbania speciosa), cowpea, soya beans, berseem, peas, groundnuts, lupins and similar crops come under this category. Certain non-legumes are also used as green manure crops and these add mainly organic matter to the soil, and little or no nitrogen. For maintaining organic matter content of the soil, the quantity or bulk is of great importance and this can often be supplied more efficiently by such crops as rye, millets, buckwheat, maize, certain oil seed crops, and a variety of weeds.

Fig. 45. A good stand of leguminous green manure crop. Sunnhemp (Crotalaria juncea) yields up to 6000 kg of green matter per ha. Green manure crops should be ploughed under the soil in the green condition before it starts to become fibrous and be forethe floweringstage. Addition of 50 to 60 kg of nitrogen per ha can be made by such manuring.

The benefits derived from green manuring can be listed as follows :

1. Organic matter is added to the soil, the decomposition of which stimulates micro-organic activity.
2. Organic nitrogen is added to the soil and is slowly released with progressive decomposition.
3. Release and mobilization of inorganic nutrients in the soil.
4. Improvement of the tilth and structure of the soil.
5. Hard subsoils are loosened with added organic matter and

permeability improved.

6. Improvement of aggregation of surface soil particles checks erosion.

Green manures are mainly used under Indian conditions by ploughing under the soil prior to planting sugarcane, paddy, maize and wheat. The usual rate of application of green manure ranges between 3000 and 7000 lbs per acre, and in the case of paddy application of green manure produces markedly improved yields compared to application of decomposed manures like composts or farmyard manure at the same level of organic matter supply.

ORGANIC CONCENTRATES

These are organic materials with a relatively higher content of plant nutrients compared to the bulky organic manures like composts, farm manure or green manures. These include the variety of oilcakes, guano, blood meal, activated sludge, etc. Oil cakes contain on an average 6 to 8% nitrogen which is readily available to crops. The high value of edible oilcakes like gingili and groundnut cakes as cattle-feed, which is in short supply, has encouraged a more extensive use of the cakes from other varieties of non-edible oil-seeds like safflower, castor, and neem as manures.

NON-SYNTHETIC INORGANIC MATERIALS

The non-synthetic inorganic materials may be considered as falling into two groups: (1) those obtained from natural salt deposits such as Chilean deposits of sodium nitrate and localised deposits of potassium nitrate, and (2) those secured as a by-product like ammonium sulphate and obtained in the process of manufacture of coke from coal. A large amount of ammonium sulphate is thus obtained in industrial countries like the United States from the various coke and gas plants.

SYNTHETIC NITROGEN MATERIALS

There are four groups of nitrogenous fertilizers depending upon the form in which nitrogen occurs, and they are produced by chemical nitrogen fixing process. These are:

(1) Nitrate fertilizers represented by sodium nitrate, calcium nitrate and potassium nitrate.

(2) Ammonium fertilizers represented by ammonium sulphate, ammonium chloride and ammonia solution.

(3) Amide or organic forms like calcium cyanamide, urea and urea-formaldehyde.

(4) Combined ammonia-nitrate fertilizers, like ammonium nitrate, calcium-ammonium nitrate and ammonium sulphate-nitrate.

TABLE XXIII

NUTRIENT CONTENT OF SOME IMPORTANT NITROGENOUS FERTILIZERS

Nitrogenous Fertilizers	*N content in %*
Ammonium sulphate	20.5
Ammonium sulphate-nitrate	26.0
Urea	46.0
Calcium ammonium nitrate	20.5
Ammonium chloride	26.0
Ammonium nitrate	33.5
Chilean (sodium) nitrate	16.0
Calcium cyanamide	21.0
Anhydrous ammonia	82.0

RELATIVE MERITS OF DIFFERENT FORMS OF NITROGEN

The question of the relative efficiency of nitrogenous fertilizers in the different forms frequently arises, as also the choice of the fertilizer forms best suited to different crops. It is difficult to decide these questions purely on the merits of each form of nitrogen but a complex of factors including the temperature and moisture conditions, soil reaction, leaching, kind of crop, and the time and method of application requires to be considered before an answer can be found. Further, the actual trials with the different kinds of fertilizers under field conditions can only give the answer in a reliable manner.

However, certain basic properties with regard to each of the nitrogen forms which are important are worth considering.

Nitrate fertilizers are readily soluble in water and in this form nitrogen is quickly utilized by most crops. Nitrates are easily leached from the soil by rains because of their high solubility and

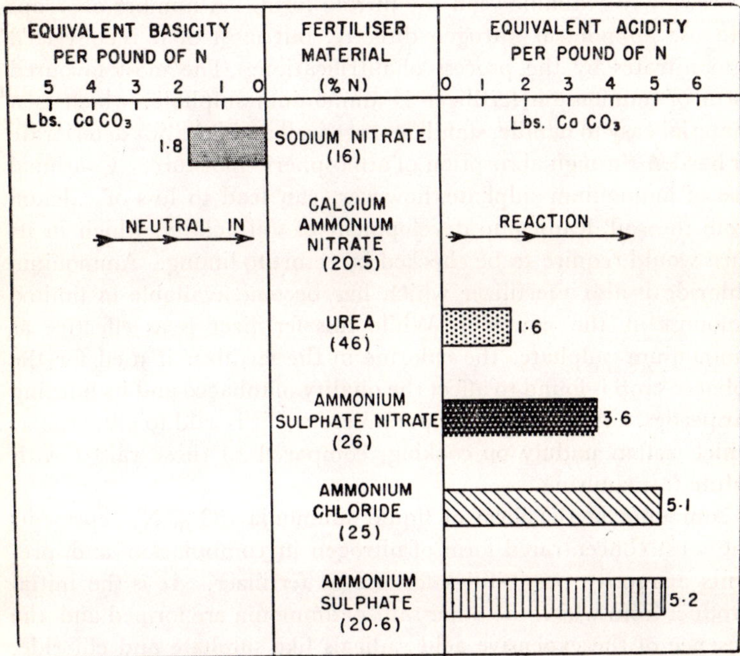

Fig. 46. The common nitrogenous fertilisers and their basicity or acidity.

Note: (i) Equivalent basicity refers to the pounds of pure calcium carbonate equivalent to the alkalinity produced in the soil from the quantity of fertiliser indicated.

Note: (2) Equivalent acidity refers to the pounds of pure calcium carbonate required to neutralise the acids produced in the soil from the quantity of fertiliser indicated.

the fact that they are not fixed or held in the soil to any appreciable extent. Nitrate fertilizers are superior to other nitrogenous fertilizers for soils with an alkaline reaction. In the use of sodium nitrate as a fertilizer, one runs the risk of adding large amounts of sodium to the soil, with its attendant disadvantages if the fertilizer is continuously applied to the soil over long periods of time.

Ammonium fertilizers are soluble in water, but the ammonia in them is not leached out so readily as nitrates, because it can be absorbed and retained in the soil. The ammonia form is, however, more slowly available than the nitrate form. A number of crops can use ammonical nitrogen directly, but most of it is converted into nitrates by the process of nitrification. The most favoured form of ammonium fertilizer is ammonium sulphate, which is a material easy to handle, stands storage well and does not deteriorate or harden through absorption of atmospheric moisture. Continual use of ammonium sulphate, however, can lead to loss of calcium from the soil, leading to development of soil acidity, which in its turn would require to be checked by resort to liming. Ammonium chloride is also a fertilizer which has become available in limited amounts in the country. While this fertilizer is as effective as ammonium sulphate, the chlorine in the fertilizer if used for the tobacco crop is found to affect the quality of tobacco and its burning properties. Also, if used for potato crops, it is said to give tubers which soften unduly on cooking, compared to those raised with sulphate manuring.

Ammonia in the form of liquid ammonia (82% N) represents the most concentrated form of nitrogen in combination and presents attractive possibilities for use as fertilizer. It is the initial product from which the other salts of ammonia are formed and the absence of the expensive acid radicals like sulphate and chloride, to form salts, makes this material much less expensive as compared to the salts. Both from the point of bulk for unit of nitrogen content and its cost, ammonia scores over the other compounds. However, the difficulties of transportation, handling, and application in the field have stood in the way of popularization of this material as a fertilizer in India. In the United States, these difficulties have largely been overcome by organisational methods, and custom application of liquid ammonia is a regular feature over most parts of that country. In our country, pioneering work has been carried out in the Mysore Department of Agriculture[1] and appliances

[1]S. V. Govinda Rajan and B. V. Venkata Rao, *Jour. Ind. Soc. Soil Sci.*, Vol. 5, No. 3, 1957.

have been devised for the application of ammonia under field conditions to common crops like sugarcane, millets and paddy. With regard to paddy, the application of nitrogen is made through the irrigation waters and a variety of trials, that have been conducted, have shown that the loss of ammonia applied to the field by

Fig. 47 Cylinder of liquid ammonia mounted on a small cart fitted with three tines enables the fertiliser material, being rapidly injected into the soil, to cover a large area of the cropped land in a reasonably short period. Studies have shown that there is little loss of ammonia from the soil through such applications. The nitrogen applied in the form of ammonia is just as effective as any other conventionally used solid fertilizers.

diffusion is negligible and the crop responses are comparable to those where fertilizers are applied in other forms at equal levels of nitrogen.

In the amide group of nitrogenous fertilizers, the nitrogen is not in a directly available form, but has to undergo chemical changes before it is converted into available forms. The important fertilizers in the group are calcium cyanamide and urea. Urea is the most concentrated form of solid nitrogenous compounds and represents 42% N in the commercial product. It has no residual harmful

Fig. 48. Application of liquid ammonia contained in steel cylinders and mounted on the farmer's bullock cart is an economical method of getting this plant nutrient into the soil.

effects in the soil which follow prolonged uses of sodium nitrate or ammonium sulphate. Urea is easily soluble in water and such solutions are used as foliar spray for fertilizing fruit trees and field crops also. Considerable economy and efficiency in the utilization of nitrogen is reported in the use of urea sprays for sugarcane.[1] Urea formaldehyde is one of the most recently developed nitrogen fertilizers, which is organic, synthetic and not water-soluble. The nitrogen in this material is released in available form slowly to provide a continual supply of nitrogen through the growing season.

WHEN AND HOW TO APPLY NITROGEN

The availability of the nitrogen applied in the form of fertilizers to crops varies with the kind and form of fertilizers. In some cases, it is immediately available to the crops and in others it is slowly avail-

[1]*Report of Fifth All India Conference of Sugarcane Research and Development Workers—1964.*

Fig. 49 Mixing liquid ammonia with irrigation waters is an economical method of applying nitrogen to growing paddy crop. The dosage is regulated by controlling a graduated valve attached to the cylinder head.

able. In any case, the effect of the nitrogen does not persist appreciably beyond the period of a cropping season. The question then arises as to what is the best method of application so that the maximum benefit of the nutrient can be derived by the crop and that, too, all through the period of growth when it is most required. In most cases, the maximum demand for nitrogen is in the early stages of the growth of the crop and this period naturally varies from crop to crop. In consideration of this, it appears useful to apply this kind of fertilizers in divided doses so as to maintain the optimum level during the period of maximum demand. Various trials have indicated that nitrogenous fertilizers in the case of paddy are given best in two divided doses, half at the time of planting the seedling, in the case of transplanted paddy, and the remaining part at the tillering stage. Application of another dose of nitrogen at a later period, say, at the time of flowering, has not proved to be an advantage. In the case of sugarcane, the application is best made in divided doses, which may be on three or four occasions ranging between the period of planting up to final earthing-up stage. The form in which nitrogen is given in these applications also appears to be important in giving increased yields, since combinations of organic and inorganic forms in differing proportions (half organic plus half inorganic, or one-third organic plus two-thirds inorganic,

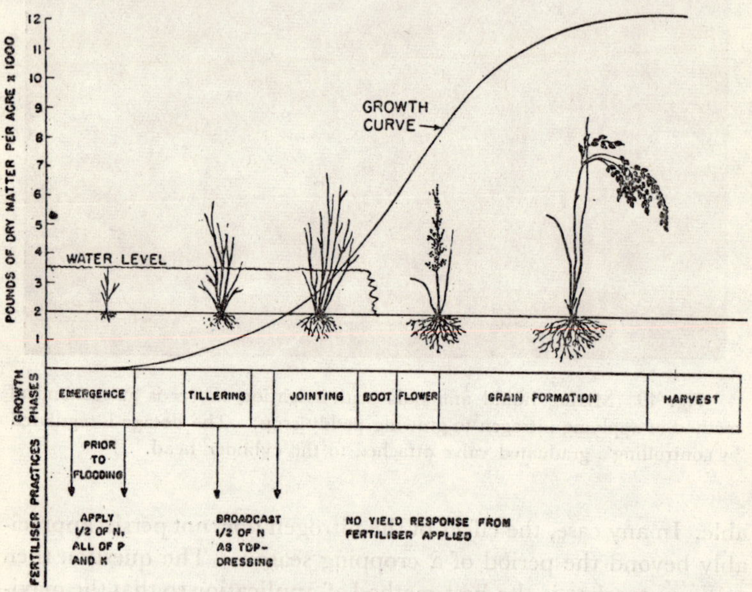

Fig. 50. Growth of rice in relation to fertiliser needs. The supply of fertilizers to crops during the periods when the nutrients are most effectively utilised by the crops ensure efficient use of the fertilizer. Application of fertilizers at wrong periods, or when crops do not utilise them, lead to wastage.

etc.) appear to be more effective than applications in purely organic or inorganic forms.

Similarly, with cotton also, field trials[1] have shown that split doses of nitrogen, part given at sowing, second part given at interculture, and the third dose at flowering stage, have given the highest response.

Placement of nitrogenous fertilizers at depth within the soil ensures that the loss of nitrogen is minimised. In the case of paddy,

[1] *Reports—I.C.A.R. Model Agronomic Trials*, 1960.

trials in India and other countries have shown that the fertilizers, in the form of pellets or mixed with clay and made into clay balls, when pushed under the root zone in the puddled field, have given better response than the surface applications.

PHOSPHATIC FERTILIZERS

In the case of phosphate supply in soils, there are no natural processes whereby this nutrient content can be increased appreciably once it is depleted. The phosphorus problem, therefore, differs from the nitrogen problem in that we cannot obtain this element from the air, but must eventually obtain it from elsewhere and supply it to the soil. In order to increase the phosphorus content of soils, fertilizers containing this element have to be applied and most of them are derived from phosphate rocks, of which deposits occur in different parts of the world.

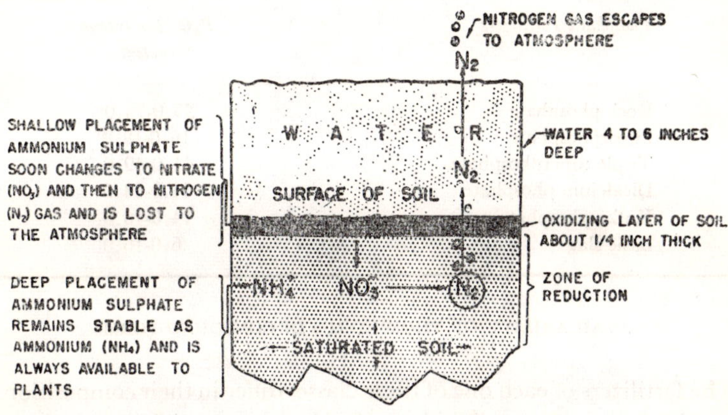

Fig. 51. Effect of placement of nitrogenous fertilizers under puddled soil conditions. Deep placement is preferable under such conditions.

FORMS OF PHOSPHATIC FERTILIZERS

The principal sources of phosphorus are the natural deposits of phosphorus-bearing rocks, iron ores and animal bones. The phosphatic fertilizers prepared from these materials may be classified into the following four categories :

(i) Natural phosphate including ground rock phosphates and bone meal.

(ii) Treated natural phosphate such as bone ash, superphosphates, calcined phosphates and metaphosphates.

(iii) By-product phosphates such as basic slag.

(iv) The chemical phosphatic compounds such as ammoniated and nitrated superphosphates, potassium phosphate, ammonium phosphate and similar compounds.

A list of the more important phosphatic fertilizers along with their nutrient contents is given in Table XXIV.

TABLE XXIV

NUTRIENT CONTENT IN PHOSPHATIC FERTILIZERS

Phosphatic Fertilizers	*P_2O_5 percentage content*
Rock phosphate	25.0-36.0
Superphosphate	16.0-18.0
Triple superphosphate	44.0-49.0
Dicalcium phosphate	34.0-38.0
Kotka phosphate	14.0-16.0
Basic slag	6.0-16.0

AVAILABILITY OF PHOSPHORUS IN COMPOUND FORMS

The fertilizers of each one of these classes differ in their composition, consistency and availability of phosphorus. The compound form in which phosphorus occurs determines the water solubility of the product and consequently its availability to the crop. Tricalcium phosphate and silico-phosphates are the least soluble of these compounds and hence phosphorus in this form is unavailable to plants unless the material is treated and the phosphate solubilized. Dicalcium phosphate is soluble in citric acid or ammonium citrate solutions and this form is considered gradually available to plants. Monocalcium phosphate and compounds of similar composition like ammonium phosphates are soluble in water and hence considered readily available to plants. Water solubility and solubility in ammonium citrate and citric acid are the criteria for assessing the availability of phosphorus in phosphatic fertilizers, and thus used to differentiate between soluble and insoluble, and hence non-available, forms of phosphorus.

SOURCES OF PHOSPHATES

Rock phosphate occurs naturally as deposits in the United States, Russia, Tunisia, Morocco and other countries. The largest known deposits are in the United States where it occurs as fluorapatite, which is a compound of calcium fluoride and phosphorus $Ca_{10}F_2(PO_4)_6$, which formula is sometimes written as $9CaO.3P_2O_5.CaF_2$. In other places it occurs as tricalcium phosphate $Ca_3(PO_4)_2$. Rock phosphate, on account of its insolubility, is rarely used as a fertilizer, and even then only after it is ground finely so that it can react better with the soil. A particularly soft variety of rock phosphate mined in Tunisia is ground to 300 mesh and marketed under the name "Hyperphosphate." This material contains 36% total P_2O_5 of which about 9% is readily available.

BONE MEALS

The bones of dead animals contain phosphorus mainly as tricalcium phosphate and these are treated to increase the availability of the plant nutrient. Various kinds of bone-meal are available, the principal ones being raw bone-meal, bone flower, and steamed bone-meal. The fine grinding of the bones, and the treatment with steam under pressure which serves to remove the fat, render the phosphorus into a more available condition. Bone-meal is more expensive compared to rock or superphosphate and is valued for application to acid soils for paddy and other crops.

Superphosphate is the most important source of available phosphoric acid and is prepared by treating finely powdered rock phosphate with an almost equal weight of sulphuric acid. The resulting product which has 18 to 20% of phosphoric acid (P_2O_5) is a monocalcium phosphate which accounts for most of the water soluble phosphate, together with small quantities of dicalcium phosphate, unreacted tricalcium phosphate and calcium sulphate. The water soluble P_2O_5 content ranges between 16% and 18% with another 2% as citric acid soluble P_2O_5.

By treatment of rock phosphate with phosphoric acid, higher grade superphosphate, containing 40-50% available P_2O_5 and commonly referred to as Triple Superphosphates are prepared. They contain very little calcium sulphate and are well suited for use in the production of concentrated fertilizer mixtures. The phosphoric acid in this is just as effective as the ordinary grades of superphosphate.

Basic slag is a by-product of the iron industry. The phosphorus in iron ores, which is undesirable in making steel, is removed by oxidation and the oxidised phosphorus allowed to react with the lime material used for lining the steel converters. The slag or scum which rises to the surface of the molten iron is drawn off, cooled and ground to a fine consistency. The phosphorus in the slag is soluble in citric acid and is considered available to crops. The slag obtained in the steel factories in India is, however, low in phosphorus content.

Of the other phosphate-carrying fertilizers, ammonium phosphate is prepared by combining ammonia with phosphoric acid (H_3PO_4) instead of with sulphuric acid to furnish a material with high plant food content. High grade compound fertilizers are prepared by combining potash salts with ammonium phosphate so that all the three major plant nutrients are had in the same material. Ammoniated superphosphate contains about 3% of nitrogen and is prepared by treating superphosphate with anhydrous or aqueous ammonia. The free acid in the superphosphate reacts with the ammonia, and the product besides containing the above quantity of nitrogen, without increase in bulk, possesses good storage properties consequent on the neutralization of the free acid.

AVAILABILITY OF PHOSPHORUS IN DIFFERENT FERTILIZERS

The phosphorus compounds in phosphatic fertilizers have been grouped on the basis of their ease of solubility as follows: (i) the water soluble, (ii) the citric acid or citrate soluble, and (iii) the insoluble. The compounds falling in the first two groups make the available phosphoric acid content of any particular fertilizer, and the insoluble compounds are considered unavailable. Monocalcium phosphate (from superphosphate) and the phosphate compounds of ammonia and potassium are readily soluble in water and hence considered available. Dicalcium phosphate and the phosphate in calcium metaphosphate and steamed bone flour are soluble in neutral ammonium citrate, and that in basic slag is soluble in citric acid. These soluble constituents are also considered available while the phosphorus in rock phosphate and bonemeal is not soluble in these solvents and hence considered unavailable.

The availability of phosphorus to crops applied in the form of fertilizers is, however, determined by a number of factors associated with soil conditions. The amount that will become available for use by a crop will depend upon the type of soil, its pH, the lime content, organic matter content, the kind of plant, seasonal conditions, and also upon the kind and amount of fertilizer applied.

FIXATION AND REVERSION OF PHOSPHATES IN SOILS

One of the problems in the use of phosphate fertilizers is the "fixation" of the soluble forms in the fertilizers by the soil into insoluble and unavailable forms. The mechanism of this fixation process has been the subject of a great deal of study by soil chemists. The fixation is partly due to the reaction of the soluble forms with constituents in the soil and their conversion into insoluble forms of calcium, iron and aluminium phosphates. Under acid conditions, the reaction is largely to form the insoluble iron or aluminium phosphates, and under alkaline conditions to "revert" to insoluble calcium phosphate forms. Also, the unavailability of applied phosphorus may be the result of a complex adsorption process, by which the phosphate radical becomes more or less firmly attached to the soil complex. Soils vary a great deal in their capacity to fix phosphate and this is related mainly to the soil pH, the organic matter content, the amount of soluble calcium and magnesium and the quantity of iron and aluminium in reactive forms.

PLACEMENT OF PHOSPHATE FERTILIZERS

Fixation of phosphate in soils is beneficial in that they do not then leach out of the soil, but the process is detrimental to the crops as they are not available to the crop and the nutrients do not move down into the soil. As a result, phosphates applied as top dressings become fixed at the soil surface and their movement to the root zone

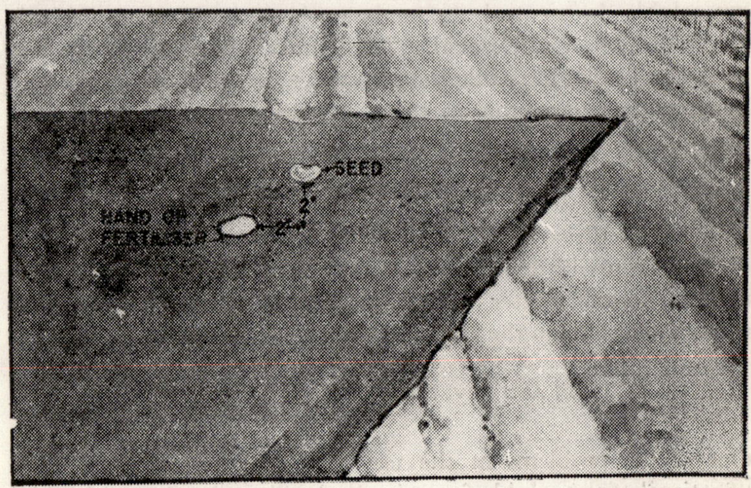

Fig. 52. A cross-section showing optimum locationing of fertilizer in relation to the seeds sown in the field. Economy in fertilizer application and better utilization of the nutrients can be ensured through such a placement.

is slowed down. To avoid such fixation losses, the procedure is to apply the phosphate at a depth near the root zone. Placement of phosphatic fertilizers in bands at depths ranging from 1½" to 2" below the seed row or the plant row and to one side of the roots results in more efficient use of the applied fertilizer.

Fixation of phosphates is much more of a problem in acid soils compared to those that have been limed to a pH value approaching neutrality. One effect of the lime is to reduce the solubility of iron and aluminium compounds in such soils, with the result that the phosphates would be more likely to be precipitated as dicalcium phosphates which are relatively more available than the insoluble iron phosphates. Various devices to place phosphatic fertilizers in depth under the soil surface have been devised, and one of these is illustrated in Fig. 52. Diagrammatic illustration of the placement of phosphates under the soil surface in relation to the sown seeds is also illustrated in Fig. 53.

INCREASING THE AVAILABILITY OF APPLIED PHOSPHATES

The processes by which relatively insoluble compounds in soil become available for plant use are not clearly understood. But it is generally believed that the various acids, organic and inorganic, that are produced in the soil, e.g. carbonic acid, nitrous acid, etc. have a part in solubilizing fixed phosphorus. By inducing the production of such acids in the soil by adding organic matter, the availability of phosphates can be increased and it is this possibility that is utilized in the application of decomposable organic matter to soils prior to cropping. An additional benefit that is likely to result from such practices is the conversion of part of the

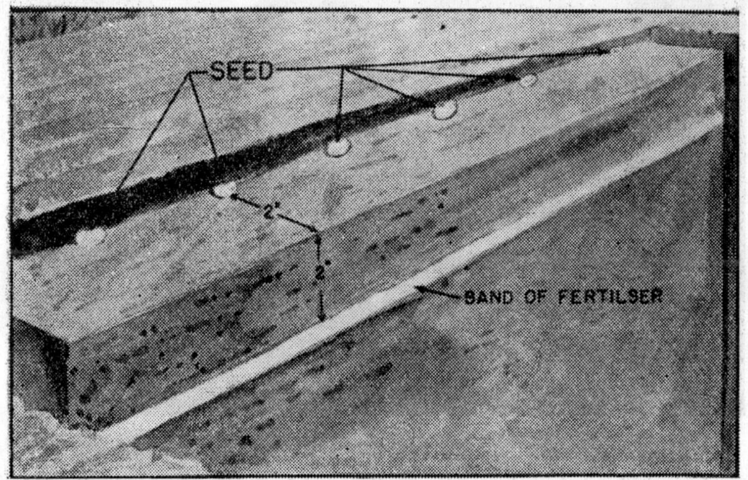

Fig. 53. A oblique section showing positioning of fertilizers at optimum depth below the soil surface in relation to the seed. This enables the best utilization of plant nutrients contained in a fertilizers.

phosphates into organic forms which do not get fixed and the gradual release of the inorganic forms through mineralization of the organic forms which can become steadily available to the growing crop. Further, the growth of leguminous crops, after application of phosphatic fertilizers, doubly benefits by improving the growth of such crops, because of better utilization of phosphorus, and the return of the green matter to the soils releases the phosphorus in the soil in available forms to the succeeding crop. In any case, under acid conditions of soil, liming helps to bring about a solubility of phosphate compounds, but overliming causes a reversion of soluble phosphates into insoluble tricalcium phosphate forms, which should be avoided.

Fig. 54. A drill that places the fertiliser approximately 2 inches to one side and two inches below the seed. Seeding and fertilising rates are adjustable. The fertiliser hopper is in front and the seed hopper is in the rear; both serve as seat for the operator.

POTASH FERTILIZES

The supply of potassium in soils is frequently much greater than the total supply of either nitrogen or phosphorus. Sandy soils contain more potassium compared to clay soils, and the potassium is released by the solubilization from a variety of minerals like feldspars, micas, etc. The problem in most soils is the question of available potassium, which is usually low, while the total reserves and supplies in the soils are much greater. The demand for potassium also varies a great deal from crop to crop and it is, therefore, necessary to supply soluble forms of potassium in the shape of potash fertilizers.

The chief potash fertilizers are the following: (i) muriate of potash (KCl) (ii) sulphate of potash (K_2SO_4); and (iii) sulphate of potash-magnesia ($K_2SO_4 . MgSO_4$)

Potash fertilizers are mainly obtained on extraction from surface deposits of potash-rich minerals and rocks, which are distributed in many countries, and the underground deposits of potash salts that occur in Germany, France, Poland and the U.S.A. Besides these large deposits, potash is also extracted from the saline waters of salt lakes, from wood ashes and from the by-products of various industrial wastes.

The principal potash fertilizers and their content of K_2O are given in Table XXV.

TABLE XXV

NUTRIENT CONTENT IN POTASSIUM FERTILIZERS

Potassium Fertilizers	*K_2O Content %*
Potassium chloride (muriate of potash)	50.0-63.0
Potassium sulphate (sulphate of potash)	48.0-52.0
Kainite	10.0-20.0

All potash fertilizers are easily soluble in water and hence they can most effectively be broadcast in the field. However, on clay soils, there is a certain amount of temporary "fixation" of potassium by the soil colloids, and hence it is preferable to apply potassium fertilizers on such soils by placement in bands. In general, the fixation of potash is not so ready as phosphates, but is more readily fixed than most nitrogen salts. The potassium ion, on being absorbed by the soil colloids, displaces some other ion such as calcium, magnesium or sodium. Plants are found to be able to utilize the adsorbed potassium which is exchangeable in nature.

In the case of crops which are affected by chlorine in fertilizers, like tobacco and potatoes, potash in the form of sulphate is preferable to the muriate form. The chloride fertilizer is found to affect the burning quality of tobacco and, in the case of potato, its cooking properties and, hence, the sulphate form of this fertilizer is better than the chloride form for these crops.

MIXED FERTILIZERS

The mineral nutrient requirement of different crops is variable as is the capacity of soils to supply these nutrients. In the application of fertilizers to meet the varying requirements of the soils and the crops, different kinds and proportions of fertilizer in different propor-

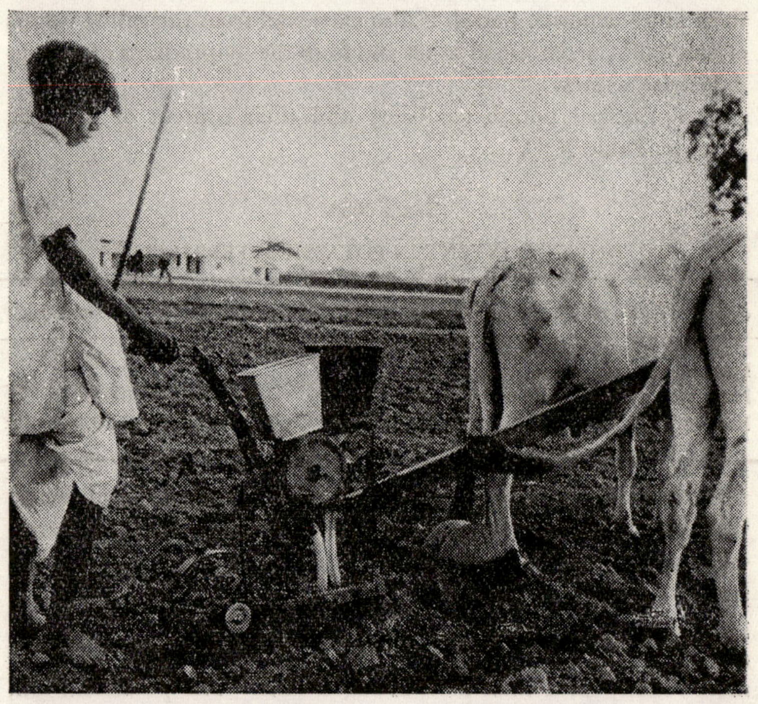

Fig. 55. A one-row seed-cum-fertiliser drill that places the seed at the proper depth and the fertiliser approximately 2 inches to one side and 2 inches below the level of the seed. The depth of seeding is adjustable.

tions need to be applied. This object is achieved by mixing different kinds of fertilizer to supply the required amounts of nutrients. Straight fertilizers are mixed to obtain either two or three nutrient elements together so that problems of transport and application are minimised. A mixture of fertilizers thus has a definite proportion or formula which indicates the amounts or proportions in which the different nutrients are present in the mixture. The use of mixed fertilizers ensures that a properly balanced fertilizer for a particular cropping condition is obtained and the difficulties and doubts about mixing on each and every occasion are avoided. Different grades of mixed fertilizers are readily available from dealers. The farmer is thus enabled to choose the required mixture.

The practice in vogue to represent the composition of a fertilizer mixture is by a formula which indicates in proper sequence the amount of N, P_2O_5 and K_2O and their respective proportions. Thus, a formula of 5 : 10 : 5 for a fertilizer mixture indicates that 100 lbs of the mixture have 5 lbs of nitrogen, 10 lbs of phosphoric acid and 5 lbs of potash. The application of 600 lbs of such a mixture to an acre will supply 30 lbs of N, 60 lbs of P_2O_5 and 30 lbs of K_2O to supply the requirements of the particular crop for which the dosage is recommended. The mixed fertilizers are manufactured by mixing suitable ingredients to supply the required quantities of the nutrients. The form in which the nutrients are to be present, as best suited to the crop for which the mixture is intended, is also considered when mixing the fertilizers. A part or the whole of the nitrogen may be in the ammoniacal or nitrate form, or some of it may be had in the organic form also. Also, with regard to phosphoric acid, consideration would be given to the form in which it should be present, whether wholly as water soluble or partly as water soluble and the remainder as citrate soluble or slowly available phosphate form. In many cases, the compounding of the individual straight fertilizers will not give the required composition and it will be necessary to add "fillers" consisting of inert materials like sand, gypsum, etc. to make up the bulk. In any case, when concentrated fertilizer constituents are used, fillers must be used in differing proportions to get the mixture of required composition.

The following grades of mixed fertilizers are in use in India, and

the different grades are recommended for application in suitable dosage for the various crops grown in the country.

TABLE XXVI

GRADES OF MIXED FERTILIZERS IN USE IN INDIA

10-10-0	5-10-10	3-9-9
6-10-6	8-8-8	8-6-10
12-6-0	5-10-5	7-7-14
9-9-0	4-8-8	6-3-12
14-7-0	6-8-8	14-6-5
12-8-0	6-6-12	12-6-9
8-8-0	6-12-6	6-9-6
5-5-0	8-4-16	8-14-12
10-9-0	9-9-9	8-6-6

COMPOUND FERTILIZERS

The availability of compound fertilizers like the ammonium phosphates nitrophosphate and potassium nitrate which supply two plant nutrients in definite proportions and proprietary products like nitrophoska, ammophoska and others, which supply all the three nutrients in specified proportions, ensures the supply of the required nutrients in minimum bulk. However, from the point of view of cost, the compound fertilizers are generally more expensive compared to mixed fertilizers. The advantage in the case of compound fertilizers is that, besides a saving in bulk, the materials have a good keeping quality and the composition of the material is guaranteed.

TABLE XXVII

NUTRIENT CONTENT IN COMPOUND FERTILIZERS

Compound Fertilizers	*Nutrient Content*		
	$N\%$	$P_2O_5\%$	$K_2O\%$
Ammoniated superphosphate	2.0-5.0	13.0-19.0	—
Diammonium phosphate	21.0	53.0	—
Monoammonium phosphate	11.0	48.0	—
Potassium nitrate	13.0	—	44.0

SUMMARY OF METHODS FOR APPLICATION OF FERTILIZERS

Nitrogenous fertilizers are applied either as a single dose at the time of sowing or transplanting or in split doses at different stages of growth.

Phosphatic fertilizers are applied at the time of sowing or transplanting in one single dose.

Potassic fertilizers are generally applied in one single dose at the time of sowing or transplanting unless a different procedure is specially indicated by experimental results or the nature of the crop.

There are three methods of applying fertilizers in the field.

1. Broadcasting: Distributing the fertilizers uniformly over the surface of the ground. This method is generally followed where large-scale applications of fertilizers are done.
2. Placement methods: Localisation of the fertilizer by application in bands or pockets under the soil surface near the root zones of the crops. This method minimises losses of fertilizers and is more effective and can be adopted for the different fertilizer nutrients. Placement is effected at suitable depths in the soil by using proper types of equipment.
3. Spraying the fertilizer solutions over the plants: This method is generally used for application of micro-nutrients to fruit trees and also nitrogenous fertilizers like urea to field and other crops. Foliar applications of soluble phosphates have also been tried and found useful.

ECONOMICS OF FERTILIZER APPLICATION

The use of fertilizers involves an additional investment by the farmer over and above the cost involved in the other operations for raising his crops. The extent to which this input will be repaid by the extra crop produced is dependent upon a number of factors, including the nature of the crop, the quantity and kinds of fertilizer nutrients applied, the moisture supply, the period of growth, and the fertility level and other properties of the soil itself. The response of crops to fertilizer nutrient application being contingent upon so many diverse factors, it becomes difficult to predict the exact increase in yield that may be expected from a particular treatment. This information can be obtained with a certain amount of reliability only from actual field trials under varying crop and seasonal conditions. Field experiments carried out at various government experimental stations furnish a great deal of useful data on the

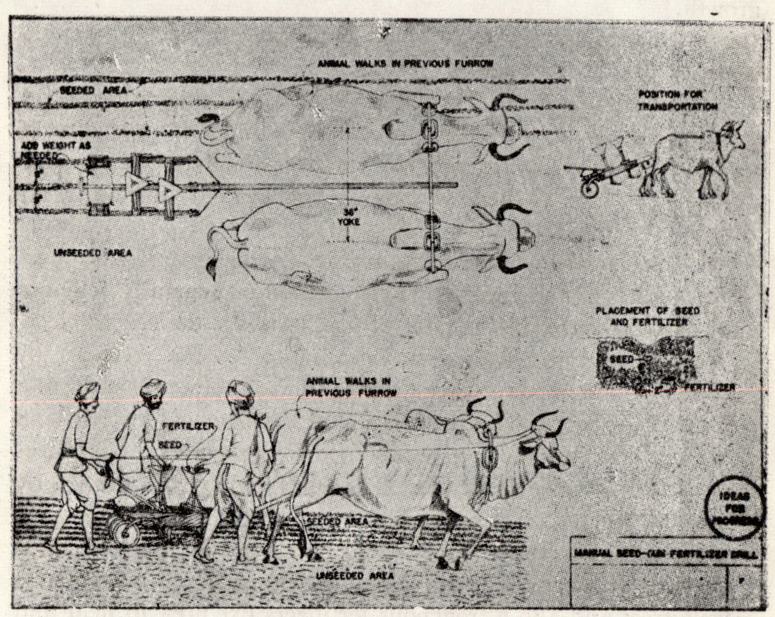

Fig. 56. A simple seed-cum-fertiliser drill, designed in India, places the seed at a uniform depth and the fertiliser approximately 2 inches to one side and 2 inches below the seed. Note also the use of bullock harness.

above, but since their applicability is limited, it is necessary that large numbers of field trials are conducted under the conditions of the farmer's operations, before reliable figures for the fertilizer responses, on which the economics of the operations depends, can be obtained.

The response of crops to fertilizer nutrient applications has been studied not only in field-scale operations but also extensively on small-scale trials in greenhouses which are relatively easier to operate. From information of the responses thus obtained, it is possible to estimate the net increase in crop and the net increase in return per unit area.

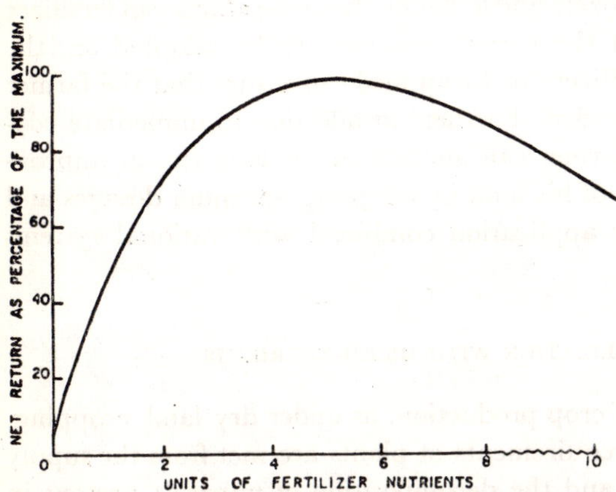

Fig. 57. Increasing doses of fertilizer nutrient showing the increase in crop yields as per cent of maximum yield in response to fertilizer nutrient additions.

The general trend of the increase in crop yields with fertilizer applications follows the curve in Fig. 57, while that of returns follows the curve in Fig. 58. The initial levels of increase in fertilizer applications tend to give higher returns, but as the levels of fertilizer applications are increased beyond certain limits, the increase in returns are reduced rapidly, beyond which it is uneconomical and also wasteful to apply more nutrients. The farmer should know the level of application which could give him the most profitable return and it is this knowledge only which can ensure that the additional input in his cultivation operations will more than adequately compensate him for the investment. In many cases, it is the wrong use of fertilizer materials and the poor returns that build up in the farmer's mind a disbelief that it is not always useful to incur additional

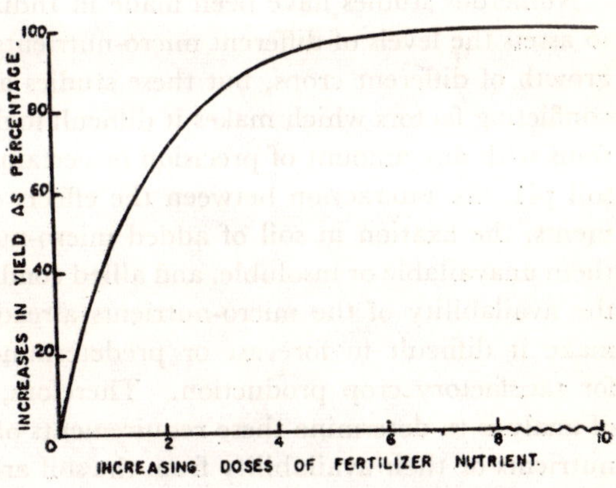

Fig. 58. Showing the net return from crop production from increasing inputs in fertilizer nutrients.

expenditure on fertilizers and a doubt about the efficacy of fertilizer use. A knowledge of the correct practices to be adopted and the right dosages of fertilizers to be applied can ensure that the farmer becomes fertilizer-minded. Further, in addition to immediate economic returns, the farmer can succeed in the long run in improving the productivity of his land by adopting optimum dosages and methods of fertilizer application combined with rational systems of crop rotation.

FERTILIZATION WITH MICRO-NUTRIENTS

Under a low level of crop production, as under dry land cropping, the micro-nutrient requirements of plants are met from the supply of organic manures and the decomposition of minerals present in the soil. Intensive crop production in irrigated agriculture and the increased crops produced through heavy manuring with the major plant nutrients make correspondingly larger demands on the available supply of micro-nutrients in the soil. Therefore, the micro-nutrients under certain circumstances may become limiting factors in increased crop production. And the intensity of these limitations will naturally vary from crop to crop because of their varying requirements for the different micro-nutrients and also their locations, as concentrations of these nutrients also vary in different areas.

Numerous studies have been made in India and other countries to assess the levels of different micro-nutrients needed for optimum growth of different crops, but these studies are beset with various conflicting factors which makes it difficult to define their concentrations with any amount of precision or certainty. The influence of soil pH, the interaction between the effects of different trace elements, the fixation in soil of added micro-nutrients which render them unavailable or insoluble, and allied conditions which influence the availability of the micro-nutrients already present in the soil make it difficult to forecast or predetermine their requirements for satisfactory crop production. Therefore, laboratory methods of analysis to determine these requirements of crops for the micro-nutrients or their availability from the soil are usually not helpful. Hence, field trials with different crops using salts of the various micro-nutrient elements, individually and in combinations, have been resorted to.

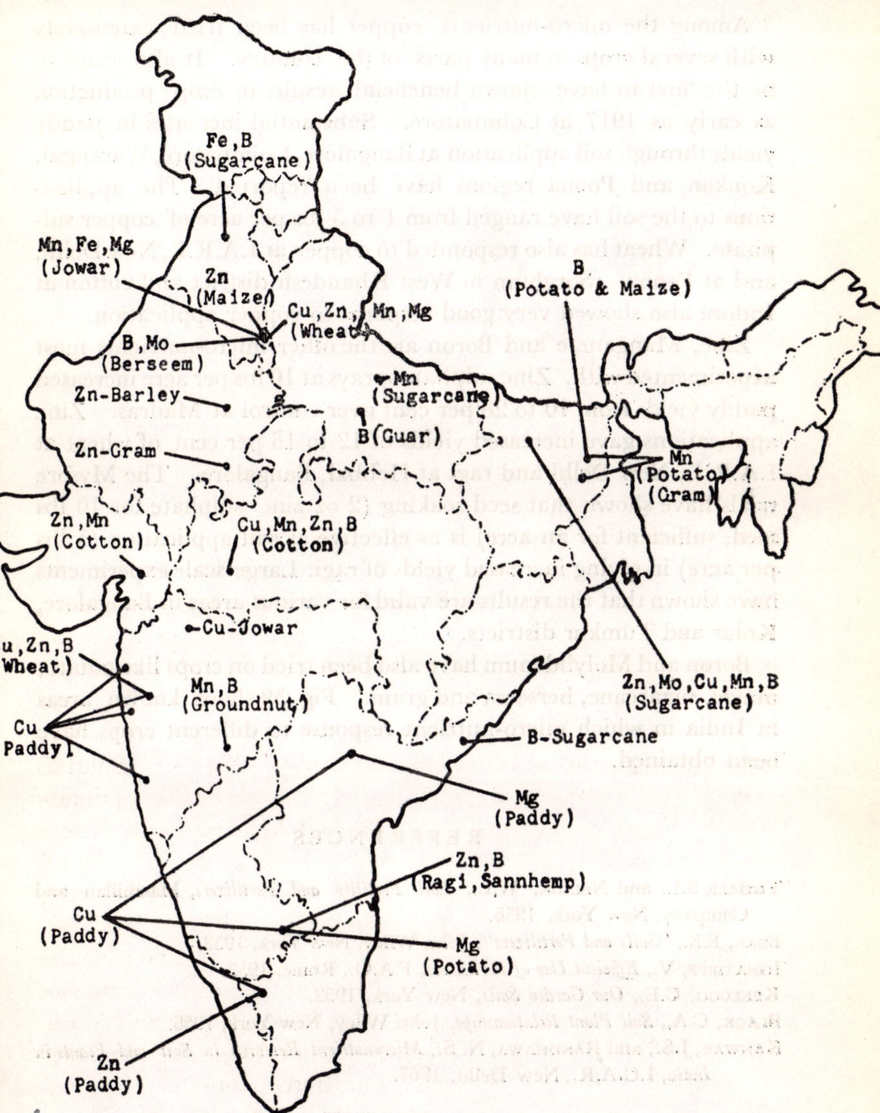

Fig. 59. Map of India showing known areas in which micronutrient deficiencies occur as revealed by the increase in growth response and, or in yields of certain crops with soil or foliar applications of Cu. Zn. Mn, B, Fe an ' Mo. Macronutrient Mg also is shown.

Among the micro-nutrients, copper has been tried extensively with several crops in many parts of the country. It also seems to be the first to have shown beneficial results in crop production as early as 1917 at Coimbatore. Substantial increases in paddy yields through soil application at Bangalore, Coimbatore, Warangal, Konkan and Poona regions have been reported. The applications to the soil have ranged from 1 to 5 lbs per acre of copper sulphate. Wheat has also responded to copper at I.A.R.I., New Delhi, and at Poona. Sorghum in West Khandesh district and cotton at Indore also showed very good response to copper application.

Zinc, Manganese and Boron are the other micro-nutrients most experimented with. Zinc sulphate sprays at 10 lbs per acre increased paddy yields from 10 to 28 per cent over control at Madras. Zinc applications gave increased yields of 12 to 16 per cent of wheat at I.A.R.I., New Delhi and ragi at Hebbal, Bangalore. The Mysore trials have shown that seed soaking (2 oz zinc sulphate for 10 lbs seed, sufficient for an acre) is as effective as soil application (5 lbs per acre) in giving increased yields of ragi. Large-scale experiments have shown that the results are valid for various areas in Bangalore, Kolar and Tumkur districts.

Boron and Molybdenum have also been tried on crops like potato, maize, sugarcane, berseem and gram. Fig. 59 shows known areas in India in which micro-nutrient response to different crops have been obtained.

REFERENCES

Tisdale S.L. and Nelson, W.L., *Soil Fertility and Fertilizers*, Macmillan and Company, New York, 1956.

Bear, F.E., "*Soils and Fertilizers*", John Wiley, New York, 1953.

Ignatieff, V., *Efficient Use of Fertilizers*, F.A.O., Rome, 1958.

Kellogg, C.E., *Our Garden Soils*, New York, 1952.

Black, C.A., *Soil Plant Relationships*, John Wiley, New York, 1956.

Kanwar, J.S., and Randhawa, N. S., *Micronutrient Research in Soil and Plants in India*, I.C.A.R., New Delhi, 1967.

CHAPTER X

LIMING AND LIMING MATERIALS

LIMING, is the addition to soil of any compound containing calcium or calcium and magnesium that is capable of reducing the acidity of the soil. The term "lime" correctly refers only to calcium oxide, CaO, but it is used almost universally to include in addition such materials as calcium hydroxide, calcium carbonate, calcium-magnesium carbonate and calcium silicate (slag).

For a number of years it has been known that soils become acid either through continual cultivation or even under natural conditions. When the acidity becomes too great or too small, certain crops fail to grow. Soil acidity then becomes a matter of very great importance to the farmer. As indicated earlier, acidity (and also alkalinity) is measured by pH values. Acid soils in India are found in the regions of high rainfall along the West Coast of India and the tracts in Northeast India, in north of West Bengal and Assam. Soils are said to be moderately acid if the pH value is below 6.0 and in the range of 5.0 to 6.0, and strongly acid if the pH value is below 5.0. Among the soils found naturally in India with very low pH, the peat soils and the "kari" soils found in Kerala show pH values below even 3.5.

Soils become acid simply because the basic elements (cations) in the soil colloids are replaced by hydrogen. When this happens,

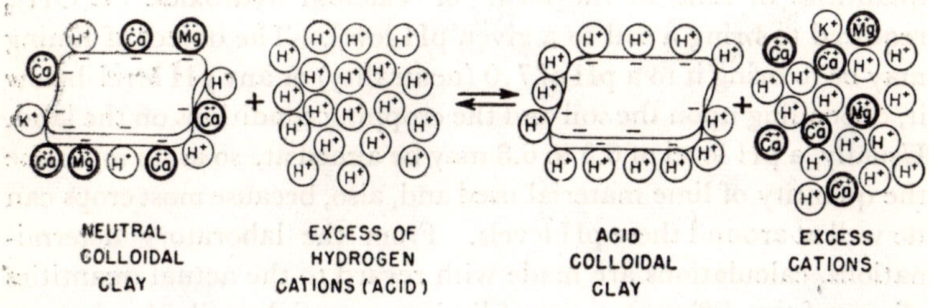

FIG. 60. Soils that are permeable and which receive more than 40 inches of annual rainfall are liable to become acid because of leaching of calcium (lime) and other bases. In this illustration, hydrogen in the soil replaces calcium and magnesium, leaving a hydrogen-saturated clay which is acid.

the soil is said to become base unsaturated. This base unsaturation is the result of being in contact with percolating waters which contain varying amounts of weak acids, resulting from decomposition of organic residues. The carbon dioxide present in soils, when dissolved in water, gives rise to weak carbonic acid, which in its turn gives the hydrogen ions which are active in replacing the bases in the soil. This continuous procession of weak acids, carbonic and also other organic acids, through the soil gradually replaces the metallic (basic) cations and the acidity of the soil increases. In cases of extreme acidity, as in the "Kari" soils, free mineral acids in the form of sulphuric acid are also released through the oxidation of sulphur compounds found in such soils in rather abnormal amounts. Even in normal soils, the continuous cropping and removal of the bases from the soil result in base deficit and a corresponding increase in soil acidity.

The important result of increase in soil acidity is the loss of calcium which is one of the essential nutrients required by plant life and which is supplied by the soil. It has been recognised by long experience that a soil having normal fertility should have enough lime supplies and lime reserves in order to maintain the yields. When supplies of lime diminish, as determined by the increase in acidity, this deficit requires to be corrected through application of adequate doses of liming materials, the correct estimation of which is made from determination of the "lime requirement" of the soil.

The lime requirement of a soil is determined by several different methods. The object of most of these methods is to determine the quantities of lime in the form of calcium hydroxide $Ca(OH)_2$ required to bring a soil to a given pH level. The object of liming may be to bring it to a pH of 7.0 (neutrality) or any pH level below it, depending upon the soil and the cropping conditions on the land. Usually, a pH level of 6.5 to 6.8 may be aimed at, so as to minimise the quantity of lime material used and, also, because most crops can do well at around these pH levels. From the laboratory determinations, calculations are made with regard to the actual quantities of any of the different types of liming material available that are to be used on the soil.

The lime requirement of a soil is related not only to the soil pH, but also to its "buffer" capacity or its capacity to absorb bases (cations) into its exchange complex. Some soils are more highly buffered than others, and the buffer or exchange capacity is related

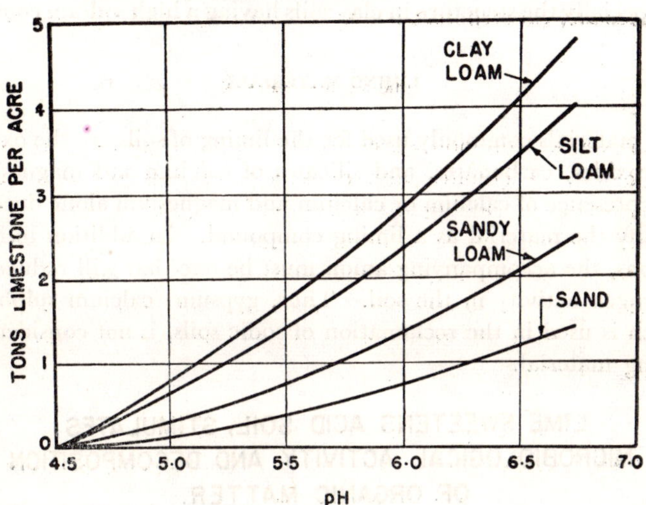

FIG. 61. The generalised relationships between the tons of limestone needed to change the pH of various soil textural classes from 4.5 to 6.7.

to the type and amount of clay and organic matter present; the larger their amount, the greater the buffer capacity. Sandy and coarse textured soils with little or no organic matter will have a low buffer capacity and so, even if very acid, will have a low lime requirement.

Liming performs several functions in the soil, and these may be briefly enumerated here.

1. Soil acidity is reduced and conditions more conducive to normal plant growth are established.

2. The essential plant nutrient element, calcium and in some cases magnesium, is also supplied through the soil.

3. Lime improves micro-biological activity directly and also by restoring conditions in soils which are satisfactory for their activities.

4. Lime releases potassium from the clay complex to be available to plants.

5. Liming checks rapid "fixation" of phosphates into non-available forms like iron and aluminium phosphates.

6. Lime offsets to a certain extent the disadvantageous conditions of excess of sodium. It improves soil structure, and especially the structure in clay soils having a high sodium content.

LIMING MATERIALS

The materials commonly used for the liming of soils are the oxides, hydroxides, carbonates, and silicates of calcium and magnesium. The presence of calcium or calcium and magnesium alone does not qualify the material as a liming compound. In addition to these cations, the accompanying anion must be one that will reduce the hydrogen activity in the soil. Thus, gypsum (calcium sulphate), which is used in the reclamation of sodic soils, is not considered a liming material.

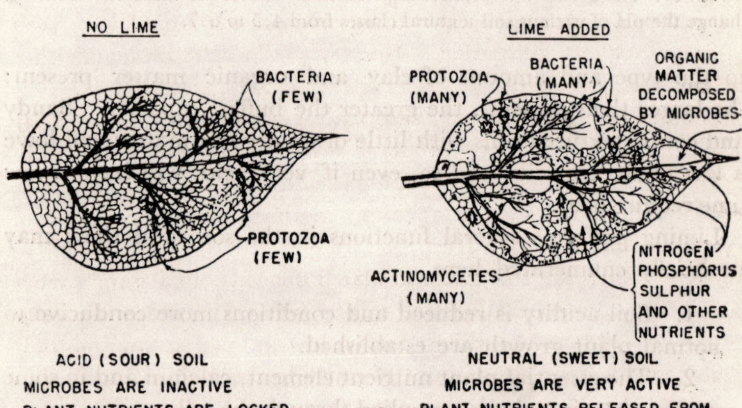

FIG. 62. Plant nutrients move from soil to plants and back to soil. In acid soils, the nutrients in the plant are not readily released back to the soil for use by other plants because decomposition is slow. Lime hastens this decomposition process and, as a result, nutrients are released faster for use by the next crop.

There are several liming materials commonly used and these include lime, shell lime, hydrated lime, ground limestone, dolomite, marl, and slag. The characteristics of each of these can be described here.

Calcium oxide. This is the only material, as indicated earlier, to which the term lime can be appropriately applied. Calcium oxide, known also as unslaked lime, burnt lime, or quick lime, is a powdery white material. It is manufactured by calcining crude limestone in the furnace or lime kilns. The purity of the burnt lime is dependent, of course, on the purity of the raw material used. When added to the soil, it acts almost immediately and, hence, when rapid results are desired from the addition of a liming compound, either this material or calcium hydroxide $Ca(OH)_2$ should be selected. In practice, this material is difficult to apply to the soil as it rapidly absorbs moisture from the atmosphere, resulting in flakes or granules.

Calcium hydroxide $Ca(OH)_2$. This is slaked or hydrated lime. It is prepared by adding limited quantities of water to calcium oxide. Much heat is evolved in the process, and the white powdery substance is collected and used. Like calcium oxide, this is a material rather difficult to handle in field applications.

Limestone and dolomite. Crystalline calcium carbonate is termed calcite or calcitic limestone, while crystalline calcium magnesium carbonate is known as dolomite in which material the calcium carbonate and magnesium carbonate occur in equimolecular proportions. When they occur in other than this proportion, the materials are said to be dolomitic limestones. Limestone is generally quarried by the open pit method, and the quality of crystalline limestones varies.

Marl. Marls are soft unconsolidated deposits of calcium carbonate. They are generally found mixed with earth and are usually quite moist. Marls are almost always low in magnesium and their value depends upon the amount of clay admixed.

Slags. There are several types of material, generally classed as slags, of which two are important agriculturally. One of them is blast furnace slag, which is a by-product in the manufacture of pig iron. As a liming material, slag behaves essentially as calcium silicate. The second material is basic slag which is a by-product of the open hearth method of making steel from pig iron, which in turn is produced from high phosphorus iron ore. The impurities in the iron, including silica and phosphorus, are melted by flux with lime and removed as basic slag. Since the phosphorus content is more useful, basic slag is applied more for its phosphorus content than as a liming material.

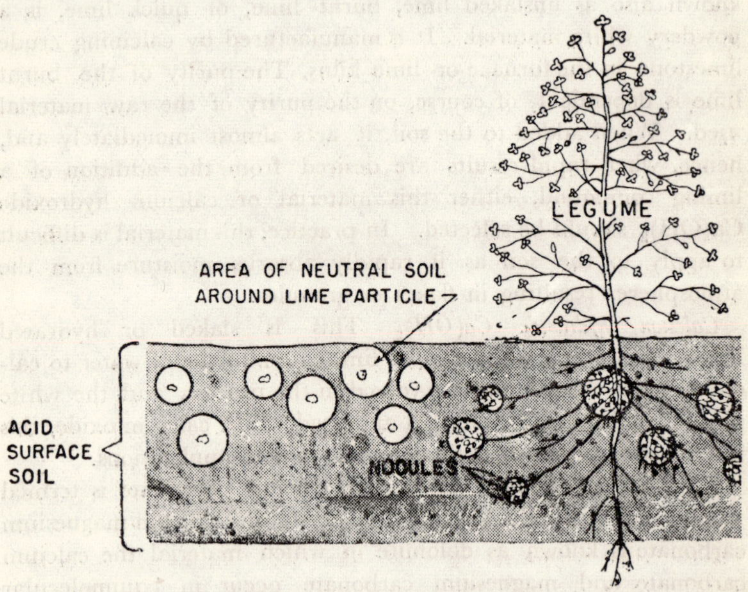

FIG. 63. How lime reacts in soil. Lime supplies calcium, sweetens soil and helps nodulation and vigorous nitrogen fixation. Every nodule contains lakhs of nitrogen-fixing bacteria.

The value of these materials in liming is based on their effectiveness as measured by their neutralising value. This effectiveness is further controlled by the degree of their fineness.

Liming materials differ markedly in their ability to neutralise acids. The value of a limestone for this purpose depends upon the quantity of acid a unit weight of the material will neutralise, and this property in turn is related to (1) the molecular composition of the material, and (2) its purity. Pure calcium carbonate is the standard against which other liming materials are measured, and the neutralising value of $CaCO_3$ is considered to be 100%. On this basis, the comparative neutralising value of pure forms of commonly used materials is as follows :

TABLE XXVIII

Material	*Neutralising value*
CaO	179 per cent
$Ca(OH)_2$	136 per cent
$CaMg(CO_3)_2$	109 per cent
$CaCO_3$	100 per cent
$CaSiO_3$	86 per cent

Use of the neutralizing value makes possible the most simple and straightforward comparison of one liming material with another relative to neutralising properties.

The neutralizing value of commonly used liming materials is as follows :

TABLE XXIX

Material	*Neutralizing value*
Ground agricultural limestone (Re-precipitated lime, lump lime)	85—100
Caustic lime	150—175
Hydrated lime (water slaked lime)	120—135
Ground dolomitic limestone	95—108
Burnt oyster shells	90—110
Marl	50— 90
Wood ashes	40— 50

Source : *The Fertilizer Hand Book,* National Plant Food Institute, Washington, 1963.

SOURCES OF LIMING MATERIALS

The most common source of lime for agricultural purposes throughout the world is high grade limestone. India, fortunately, has a plentiful supply of limestone deposits, but so far they have not been used to any extent to supply ground limestone to farmers. Other deposits of limestone occur throughout India but their exploration, analyses, and potential use must wait until a greater demand is created.

Table XXX
LOCATION AND CALCIUM CARBONATE EQUIVALENT OF LIMESTONE AVAILABLE IN INDIA*

State	*District*	*Calcium Carbonate Equivalent (per cent)*
Andhra Pradesh	Kurnool	85
	Hyderabad	—
	Visakapatnam	—
Assam	Khasi Hills	94
	Garo Hills	—
Bihar	Shahabad	87
	Hazaribagh	—
	Palamau	—
	Singhbhum	—
Gujarat	Junagadh	96
	Banaras	—
Madhya Pradesh	Jabalpur	85
	Durg	—
	Bilaspur	—
Madras	Tiruchirapalli	100
	Nellore	—
	Salem	—
	Madurai	—
	Ramanathapuram	—
Maharashtra	Yeotmal	98
Mysore	Shimoga	89
	Chitaldurg	—
	Tumkur	—
	Mysore	—
	Bijapur	—
Orissa	Sundargarh	96
	Sambalpur	—
	Koraput	—
Punjab	Patiala	95
	Ambala	—
Rajasthan	Bundi	81
	Jodhpur	—
	Udaipur	—
Uttar Pradesh	Mirzapur	—
	Dehra Dun	—
West Bengal	Darjeeling	—

* M. Chakraborti, B. Chakravarti and S. K. Mukherjee, "Liming in Crop Production in India," *I.C.A.R. Bulletin No.* 7, 1961.

FINENESS OF LIMESTONE

Molecular constitution and freedom from inert impurities are not the only properties determining the effectiveness of agricultural limestone. The degree of fineness is equally important, since the speed of their reaction with the soil depends upon the surface exposed to such a reaction. The approximate relative values of limestone of varying particle size evaluated 1, 4 and 8 years after application are given below.

Table XXXI

Size fractions	*Period after application*		*Relative values*
	1 year	*4 years*	*8 years*
Through 60 mesh	100	100	100
30-60 mesh	50	100	100
20-30 mesh	25	60	100
8-20 mesh	15	30	50
Over 8 mesh	5	15	25

SOURCE: *Bulletin No. 721*, University of Illinois.

The degree of fineness is measured by sieving. The "60 mesh" indicates that the sieve has 60 apertures to a square inch, 30 mesh has thirty apertures to a square inch, and so on.

USE OF LIME IN AGRICULTURE

The application of lime to many soils results in striking increases in plant growth when crop responses are obtained from the application of materials carrying the major plant nutrients, nitrogen, phosphorus, and potassium. It is assumed, and usually correctly, that the response was the direct result of correcting only a deficiency of one of these nutrient elements. Responses from the application of lime, however, cannot always be attributed to the plant nutrient value of calcium or magnesium, because of their influence on various other factors in the soil, as also the reaction of the soil itself.

EFFECT ON PHOSPHORUS AVAILABILITY

The relationship between the availability of phosphorus in the soil and the soil pH has been discussed earlier. At low pH values, iron and aluminium compounds present in the soil become active and react with phosphorus converting it into relatively insoluble compounds which are unavailable to plants. The addition of a liming agent, while reducing the pH, causes a precipitation of the iron and aluminium in other forms and thus increases the plant available phosphorus. Application of excessive amounts of lime, again, reduces phosphate availability because of the precipitation as calcium or magnesium phosphates. Hence a liming programme should normally be planned to keep the pH between 6.0 and 7.0 to get the maximum benefit from applied phosphate.

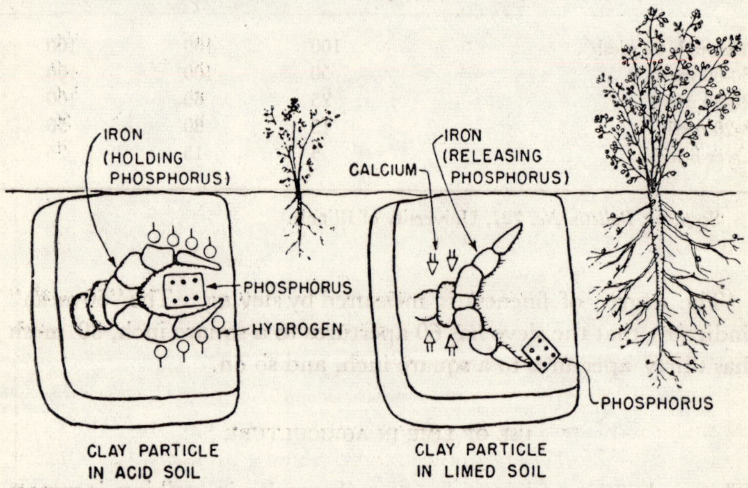

FIG. 64. Soils are acid because of excess of hydrogen. In acid soils, phosphorus is fixed by iron. When an acid soil is limed, calcium replaces hydrogen, inactivates iron, and releases phosphorus for plant growth.

MICRONUTRIENT DEFICIENCY

With the exception of molybdenum, the availability of micronutrient elements increases with a decrease in pH. The addition of excessive amounts of lime can lead to certain micronutrient deficiencies, particularly of manganese, boron and iron, and therefore should be avoided. It is a wise procedure to keep the pH at less than 7.0 while applying lime, so as to avoid causing non-availability of these micronutrients to plants.

NITRIFICATION

The organisms carrying the nitrification in soils by conversion of ammonia into nitrates require considerable amounts of active calcium. Hence the presence of adequate quantities of lime helps these organisms, and also, in general, helps microbial breakdown of soil organic matter and of plant residues, which in their turn promote microbial activity in the soil and thereby keep the soil in a healthy condition.

PHYSICAL CONDITION OF SOILS

The structure of fine textured or heavy soils can improve with the application of lime. This improvement is partly due to increase in the organic matter through breakdown of plant residues and partly because of the conversion of the clays into calcium-rich clays, which flocculate more easily and also develop good structure.

PRECAUTIONS TO BE OBSERVED IN LIME APPLICATION

While the benefits accruing to the land through the applications of lime are detailed above, certain precautions in the application of lime to soils require to be observed to prevent harmful conditions.

Lime should never be allowed to be mixed with farmyard manure, either in storing or during field application, as loss of nitrogen in gaseous form from the manure can result.

Lime should be applied to the lands one to two months prior to sowing the seeds, and applications just before sowing should be avoided to minimise possible injury to the seeds, reduction in their germination and to provide for a normalization of the conditions in the soil before cropping operations are taken up.

Lime does not move easily in the soil and, unless it is well mixed, it tends to stay where it is applied. Hence lime, after application, should preferably be mixed with the top soil through cultivation.

Indiscriminate use of lime, particularly on sandy soils, should be avoided as the disadvantageous symptoms show up more quickly than in other kinds of soils. A soil test for liming is made once in five years.

Liming is of benefit to increasing crop production, particularly of leguminous crops and those which require extra amounts of calcium for the growth and development of seeds. Soils in the coffee-growing areas in India, as in the Malnad districts of Mysore State, tend to become highly acidic due to the presence of large amounts of organic matter in these soils and the leaching action

Fig. 65. Application of lime to soils is made by broadcasting it evenly over the land after the first ploughing. Subsequent ploughing and cultivation operations enable the lime to be mixed thoroughly with the soil.

of the heavy rainfall in these regions. It is necessary to lime such soils periodically, but only to the extent of reducing the acidity to the optimum required for coffee growth and not bringing the soils to neutral pH.

BENEFIT OF LIME APPLICATION TO FIELD CROPS

Various experiments conducted in the country have shown the benefits of application of lime alone and with fertilizers supplying the major plant nutrients. While application of lime alone has shown improvement in crop yields, these increases are more marked when it is applied along with fertilizers. The enhancing effect of lime is possibly due to increase in the availability of the nutrients, particularly phosphorus and potassium, to the growing crop and thereby increasing their yield. Even with a crop like paddy which is grown over a fairly wide range of pH conditions, the application of lime alone and with phosphatic fertilizers on moderately acid soils has given increased yields. (See Table XXX)

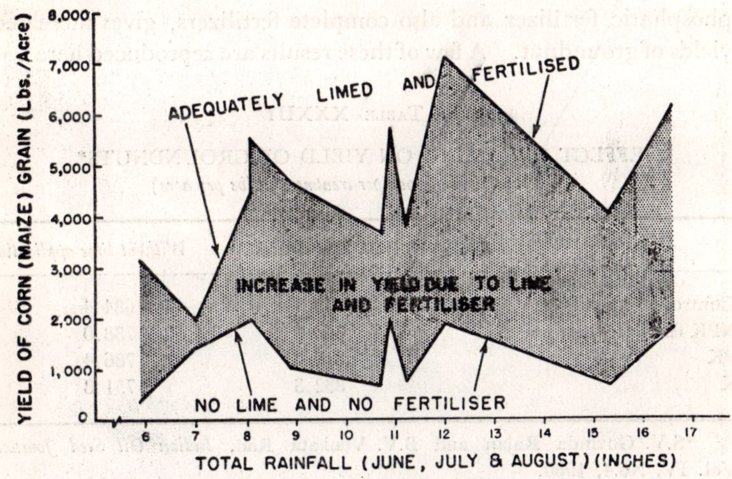

FIG. 66. Every year for 11 years, even though the summer rainfall varied from 6 to 17 inches, the yield of maize was higher when it had been adequately limed and fertilised. The average increase in yield due to lime and fertiliser was 250 per cent. (Illinois, U.S.A.)

TABLE XXXII

EFFECT OF LIME AND PHOSPHATE ON PADDY CROP*

Treatment	*Yield in lbs per acre*			
	1949-50		*1950-51*	
	Grain	*Straw*	*Grain*	*Straw*
Control (only 56 lbs bone-meal)	2299	4011	1958	3660
100 lbs lime+56 lbs "	2755	4344	2365	3679
200 lbs lime+56 lbs "	2331	4021	2248	3656

*Paddy Breeding Station, Nagenahally, Mysore.

In the above experiments, two levels at 100 and 200 lbs of lime (CaO) per acre and phosphatic manure (bone-meal) at 56 lbs per acre were used and a green manure crop raised. The soil pH was 5.6. The green manure (Sunhemp) was ploughed into the soil prior to planting paddy. The paddy crop received a uniform dose of nitrogenous fertilizers (ammonium sulphate at 112 lbs per acre). The increases in grain yield ranged up to 456 lbs per acre.

Field experiments have shown that lime alone, or added to phosphatic fertilizer and also complete fertilizers, gives increased yields of groundnut. A few of these results are reproduced here.

TABLE XXXIII

EFFECT OF LIMING ON YIELD OF GROUNDNUTS*

(*Average yield of pods per treatment in lbs per acre*)

	With lime application	*Without lime application*
Control (No fertilizers)	775.1	684.4
NPK fertilizer mixture	843.9	788.0
PK	878.4	786.4
K	832.3	751.8

*S.V. Govinda Rajan and B.V. Venkata Rao, *Indian Oil Seed Journal*, Vol. IV, No.4, 1960.

The treatments were:

Lime at ½ ton per acre

N at 10 lbs of N (as ammonium sulphate) per acre

P at 60 lbs of P_2O_5 (as superhosphate) per acre
K at 30 lbs of K_2O (as muriate of potash) per acre

Near Ranchi, (Bihar) where the average annual rainfall is 43 inches, in 1958-1960, complete fertilisers alone, lime alone, and complete fertilisers plus lime were tried in acid soils on wheat, maize, gram, and groundnut. Comparing lime alone with N+P+K fertilizers alone, lime was 77 per cent as effective on wheat; 79 per cent on maize; 98 per cent on gram; and 188 per cent as effective as fertilisers alone on groundnut (see table XXXII).

TABLE XXXIV

RESPONSE OF SELECTED CROPS TO LIME, FERTILIZERS, AND LIME PLUS FERTILISERS IN RANCHI, BIHAR*

Treatment	*Yield in pounds per acre*			
	Wheat	*Maize*	*Gram*	*Groundnut*
No lime, no fertiliser	557	594	156	550
Lime alone	700	1,000	374	1,225
N+P+K fertiliser alone	915	1,265	383	650
N+P+K fertiliser plus lime	1,276	1,815	898	1,400

*M. Chakraborty, B. Chakravarti, and S.K. Mukherjee, "Liming in Crop Production in India," *I.C.A.R. Bull. 7*, 1961.

REFERENCES

TISDALE, S.L. and NELSON, W.L., *Soil Fertility and Fertilizers,* Macmillan and Company, New York, 1956.

BUCKMAN, H.D. and BRADY, N.C., *The Nature and Properties of Soils,* Eurasia Publishing House, New Delhi, 1964.

BHAUMIK, H.D. and DONAHUE, R.L., *Soil Acidity and the Use of Lime in India,* Min. of Food and Agrl., New Delhi, 1964.

WORTHEN, L.E. and ALDRICH, R.S., *Farm Soils, Their Fertilisation and Management,* John Wiley, New York, 1956.

CHAPTER XI

SOIL TESTING IN INDIA

Soil analysis as a means of assessing the physico-chemical properties and the nutrient content in soils has been long recognised as a basis for scientific management of soils and as a rational guide for the application of manures and fertilizers for increased crop production. The traditional methods of chemical analysis are laborious and time-consuming, and as the analysis of large numbers of samples of soils involves a great deal of work and delay, methods for rapid determinations have been evolved in recent times. In preference to volumetric and gravimetric determinations, which are a feature of the traditional methods, electrometric methods using photocells have come to be adopted to speed up the process. Further, the time-consuming determination of soil bases like potassium and sodium has been replaced by light measurement techniques using flame photometers, and the same equipment can be used also to determine rapidly calcium and magnesium.

The simplification of the laboratory procedures of plant nutrient determination and the speeding up of this process have made it possible to undertake large-scale analyses of soils and thereby bring the evaluation of the qualities of the soil within the reach of the common farmer. Soil testing, as a ready means to measure the nutrient deficiencies in the soil, has come into vogue during the last two or three decades. Analysis of soils in soil testing laboratories in the United States of America as a service to the farmer has come to be developed extensively during the past twenty years or so. It has now been accepted as one of the important routine advisory services to farmers in most of the agriculturally advanced countries. It is not unusual for such a laboratory in the United States to handle 25,000 or more soil samples in a year.

Recognising the utility of the soil testing service as a means of educating the Indian farmer to recognise the properties of his soil, its potentialities and deficiencies based upon scientific examination of the soil and to guide him in the use of fertilizers, the Soil Testing Service in India was organised with the aid of the Technical Cooperation Mission of the U.S.A. Twenty-four laboratories distributed in different parts of the country were set up with equipment furnished by this mission and have been placed in charge of chemists suitably trained in the techniques of operating the various methods

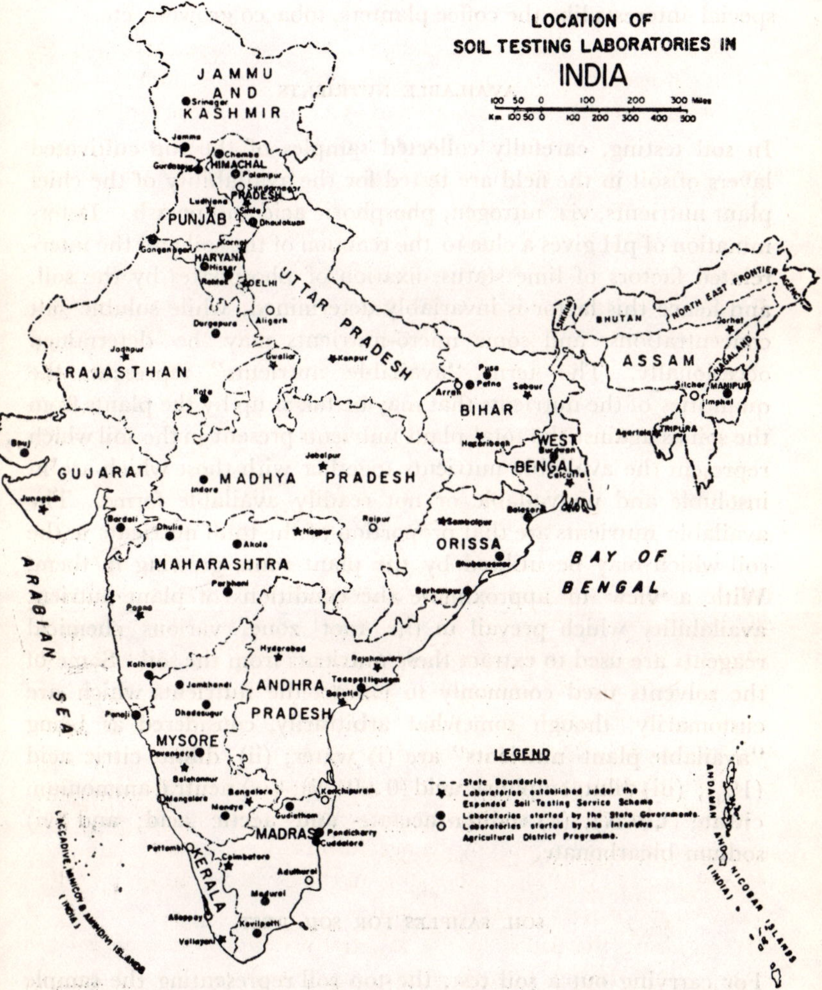

FIG. 67. Twenty-four soil-testing laboratories were started under the Expanded Soil Testing Service Scheme. Additional laboratories are being established by the State Governments and under A.I.D. programmes.

of rapid analysis. The distribution of these laboratories in India is given in Fig. 67, and the addresses of these centres are given in the Appendix. It will be seen that at least one laboratory has been set up in each state in the country and, in some cases, more than one exist, based upon the demand for such laboratories from special interests like the coffee planters, tobacco growers, etc.

AVAILABLE NUTRIENTS

In soil testing, carefully collected samples of the top cultivated layers of soil in the field are tested for the availability of the chief plant nutrients, viz. nitrogen, phosphoric acid and potash. Determination of pH gives a clue to the reaction of the soil and the inter-related factors of lime status, fixation of phosphates by the soil, and hence this factor is invariably determined, while soluble salt concentrations and some micro-nutrients may be determined occasionally. The term "available nutrients" represents the quantities of the nutrients that may be taken up by the plants from the soil as against the total plant nutrients present in the soil which represent the available nutrients together with those which are in insoluble and unavailable or not readily available forms. The available nutrients are that proportion of the total nutrients in the soil which may be utilised by the plant roots growing in them. With a view to approximate the conditions of plant nutrient availability which prevail in the root zone, various chemical reagents are used to extract these nutrients from the soil. Some of the solvents used commonly to extract the nutrients which are customarily, though somewhat arbitrarily, considered as being "available plant nutrients" are (i) water; (ii) dilute citric acid (1%); (iii) dilute sulphuric acid (0.004 N); (iv) neutral ammonium citrate (1.0 N); (v) sodium acetate and acetic acid; and (vi) sodium bicarbonate.

SOIL SAMPLES FOR SOIL TEST

For carrying out a soil test, the top soil representing the sample from a layer of the top six to nine inches, which is commonly considered as representing the plough depth, is taken. The roots of most kinds of the common field crops operate within this zone of the soil layers and hence collection of samples from this region is considered adequate. This consideration is also based on the fact that the texture, structure, moisture-holding capacity, organic matter content, microbiological activity and such factors in respect of this top soil are usually considered as important for determining the quality of the soil for crop production. The analysis of the top soil is, therefore, considered adequate to assess the fertility status of the soil and furnish the farmer with advice in regard to the application of fertilizers and also the undertaking of simple reclamation

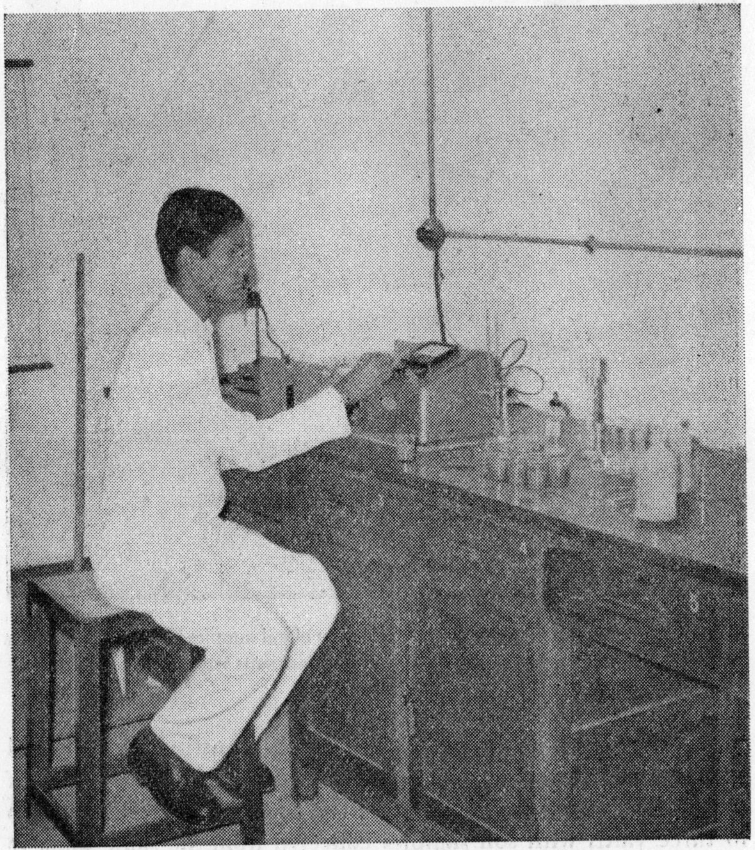

FIG. 68. Analysis in the laboratory of soil samples includes determination of pH, and available phosphorus and potassium, besides other nutrients.

or amelioration methods. Soil testing by these methods, however, is not a substitute for the detailed information that is collected by a detailed soil survey whereby information in regard to the characteristics of the soil profile, the depth of the soil, its erodability, drainage characters and such other information are collected for recommending proper land use and management. A soil test is a

Fig. 69. Shows the usual method of collection of soil samples in the field. Samples may also be tested in the field itself for pH and nutrients with rapid test kits.

means of rapidly analysing the soil for the constituents indicated earlier, the soil being taken from individual fields or a group of plots. These tests may be repeated at short intervals of, say, two to three yards with soil samples collected from the same plots to assess the effects of repeated cropping or the residual effects of fertilization.

COLLECTION OF SOIL SAMPLES

A sample of soil collected for a soil test should be representative of the field or the area selected as, otherwise, the analysis, however accurate, will not give a true picture of the conditions prevailing in the field and will not, therefore, be useful. The qualities of soils vary within fairly wide limits from place to place and hence a simple soil sample taken from one place in a field cannot be taken as representing the characters of the entire field. It is necessary to obtain a composite sample composed of several individual samples collected from a number of spots located at different places within

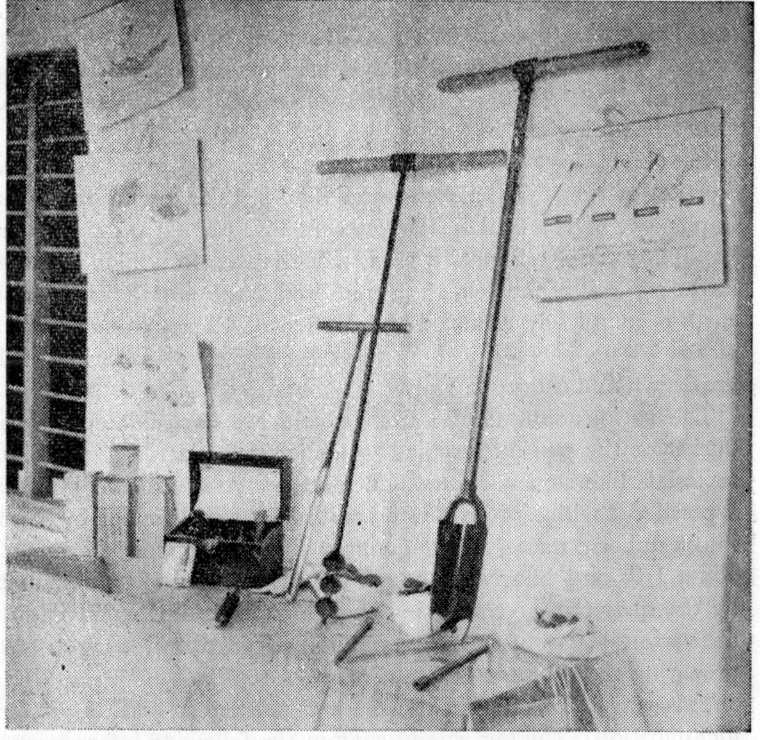

FIG. 70. Different types of augers and other instruments are used for sampling for soil test. Usually, soils up to 9 inches deep are taken as being representative of the plough furrow layer from which plants take the soil nutrients. In case of need, a second, lower than 9 inches, layer is also sampled and tested.

a field. A number of individual samples, referred to as "sub-samples," usually ranging from 10 to 16 in number are collected from different spots located at random over the selected field. These are consolidated and a composite sample representative of the whole and weighing, say, ¼ to ½ kilogram, is taken for testing.

The samples of soils collected from each random spot as indicated above should be representative of the entire depth collected, which may be 6 to 9 inches and should not be just quantities collected from the surface only. This type of representative collection is ensured by following certain techniques or using the proper types of implements. The following equipment is found useful for the collection of soil samples in the field : (i) a soil sampler which may be (a) soil boring tube, or (b) soil auger, or (c) posthole auger, or (d) a spade, or (e) a khurpi; (ii) plastic or enamelled bucket or pan, (iii) sample bags of cloth or alkathene (7″x4″); (iv) strips of paper (3″x2″); (v) strings; and (vi) information sheets—blank forms for noting soil history.

If a soil boring tube or an auger is used, it is easy to get a uniform section of the soil of the required depth. The operation of this implement in the field is illustrated in Fig. 69. If on the other hand, a spade or a khurpi is used, a V-shaped excavation is made in the top soil of the field to the required depth and the excavated earth is completely removed and discarded. Then from one side of the V-cut, slices 1 inch in thickness of the soils to be sampled are drawn and collected.

The various sub-samples from a field are consolidated in the bucket or the pan and carried to a sheltered place. Extraneous materials like stones, leaves, and roots are removed and the soil is powdered with a piece of wood and is thoroughly mixed by hand into a uniform mass. If the quantity thus obtained is in excess of about 1 lb or ½ kilogram, the quantity is reduced to this amount by the quartering technique. The illustrations in Fig. 71 depict the various stages of collection of the individual soil samples and preparing the representative composite sample.

This sample is then put in a cloth or plastic bag and tagged for identification. The tag should contain the following information : (i) Name of the farmer; (ii) Location of field (survey number) place, village, taluk, district; (iii) Date of collection; and (iv) Sample number.

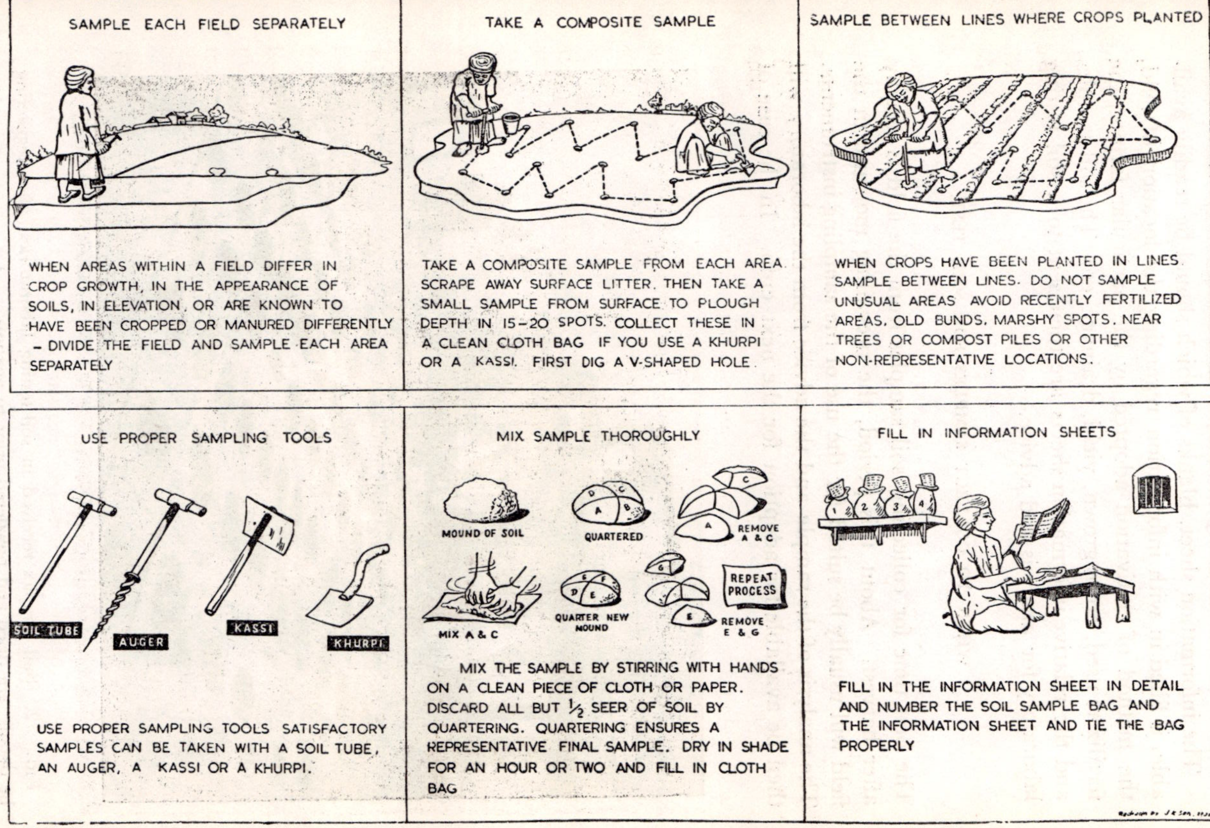

FIG. 71. Shows diagrammatically the different steps to be followed for collection of soil samples for soil test.

The bag is then tied up securely with a string.

The information sheet, blanks of which should be readily available, is filled in with information regarding the location of field, the method of cultivation adopted by the particular farmer, the fertilizers used, crops grown, yield, drainage, etc. The soil samples and the relative information sheets are then sent to a soil testing laboratory for analysis and advice.

WHEN TO COLLECT SAMPLES FOR SOIL TEST

The best time for collecting soil samples from the field is usually after harvest. About this period, the prevailing moisture in the field will usually be suited for the use of the sampling instruments. The analysis of the sample and the advice from the laboratory will then be available for adoption for the next crop. In the case of

FIG. 72. Soil samples received in soil testing laboratories, after registering he details, are dried and powdered prior to analysis.

perennial crops like orchard crops, plantation crops, etc., the sampling may be done at any time, only avoiding the period within the three months immediately after fertilizers have been applied. In cases where there is intensive cultivation or repeated cropping, as in the case of vegetable gardens adjacent to urban areas, the samples have to be collected first at harvest time and the results used either for the immediately following crop or the crop after it. Where the soil in the fields is very hard for the use of sampling instruments, sampling may be done in the ploughed fields.

RESULTS OF ANALYSIS AND RECOMMENDATIONS

The results and recommendations of soil tests may be used in the field in the same season if they are received prior to the period of application of fertilizers. If, for any reason, the recommendations are received after the period, they will be valid for the next season for the same crop. If the cropping is changed, the recommendations should be suitably altered, if necessary, with the help of the local Agricultural Extension Officer.

The Soil Test Reports indicate the level of each major plant

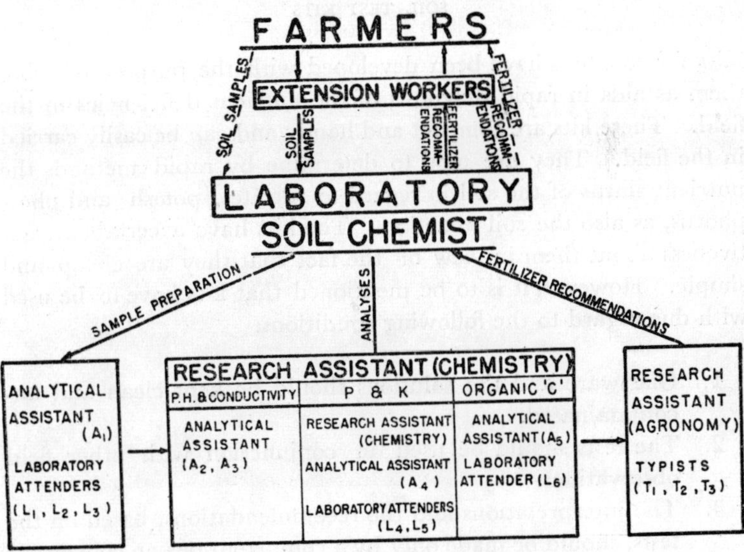

FIG. 73. A chart outlining the organisation of the soil testing service in India.

nutrient in the soil, categorised into three groups, viz. low, medium and high. The salt concentration in the soil, if any, will also be indicated. Similarly, the soil reaction or pH will be known, i.e. whether a soil is acidic, neutral or salt affected. These results give a generalised and broad indication of the nature of the soil deficiencies and they will have to be supplemented with other field information to arrive at proper conclusions. The recommendations for fertilizers are made on the basis of individual major nutrients only, viz. nitrogen, phosphoric acid and potash. The fertilizers recommended may be the ones which are easily available and popular. However, in case fertilizers, other than those suggested, are to be used, equivalent amounts of these would require to be applied to give the same dose of the recommended nutrient. The recommendation report usually indicates the equivalent of different fertilizers to supply the recommended dosages of plant nutrients. When it is desired to apply mixtures of fertilizers, equivalent quantities of the mixtures to give the recommended doses of the nutrients should be used, differences, if any, being made up by addition of suitable quantities of straight fertilizers.

SOIL TEST KITS

Kits for soil tests have been developed with the purpose of using them as aids in rapid diagnosis of the nutrient deficiencies in the field. These kits are compact and handy and can be easily carried in the field. They are used to determine by rapid methods the nutrient status of the soil in regard to nitrates, potash and phosphorus, as also the soil reaction. The kits have a certain attractiveness about them in view of the fact that they are cheap and simple. However, it is to be mentioned that kits have to be used with due regard to the following conditions:

1. Glassware and the solutions should be kept clean and uncontaminated.
2. The tests should be used in conjunction with other field observations.
3. The interpretations and the recommendations, based on the tests, should be made only by a competent person conversant with local conditions and agronomical practices.

The soil test kits, therefore, cannot be a real substitute for a proper analysis of soils in well-equipped soil testing laboratories as is imagined in popular circles. However, if one is aware of these limitations, the soil testing kit can be a very useful and cheap tool for making field recommendations by a discriminating extension worker.

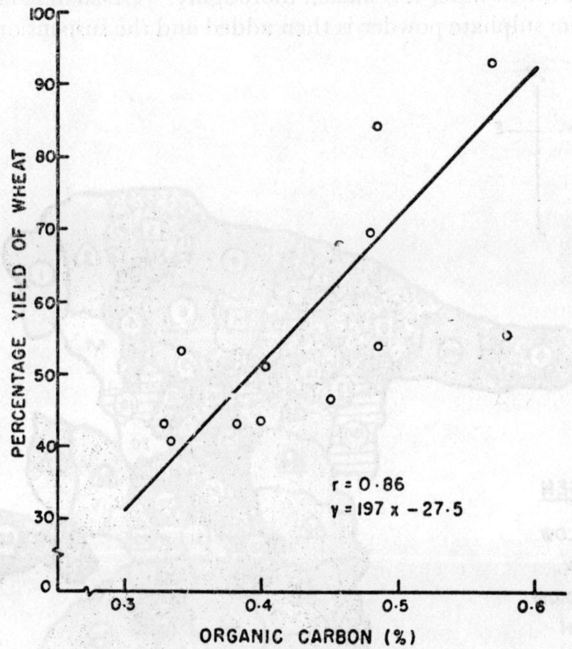

Fig. 74. Correlation between the percentage of carbon in the soil by the Walkley and Black method and the yield of wheat is highly significant. (Pot culture trials were conducted at the Indian Agricultural Research Institute New Delhi.)

Different kinds of soil test kits have been developed abroad and many types have appeared in the Indian market in recent years. Some of these are of American and others of English manufacture. While these kits have been found useful in their respective countries, their popularity in this country is limited by the need to obtain replacements of test reagents and solutions from abroad. Recently, at the Indian Agricultural Research Institute, New Delhi, a soil test kit to suit Indian conditions has been developed. Prototypes of these soil test kits are now manufacturd by private agencies in the country.

Some of the principal features of the I.A.R.I. Soil Test Kit are given below.

1. *Soil Reaction* (pH)[1] is determined by taking a quarter teaspoonful of powdered top soil in a glass test tube and after adding 5 cc of distilled water it is shaken thoroughly. One half to one gram of barium sulphate powder is then added and the suspension is kept

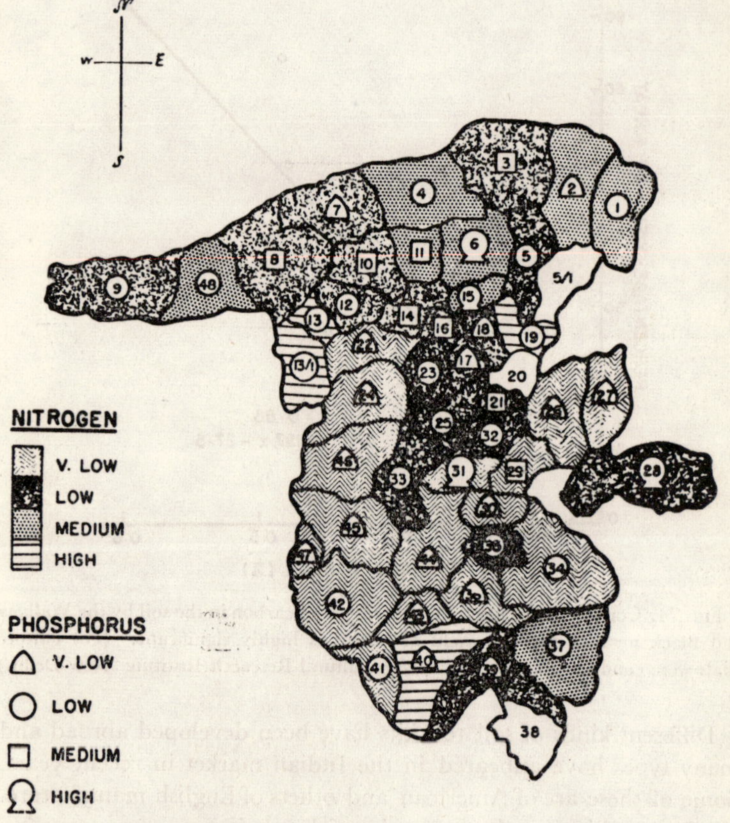

Fig. 75. A soil fertility map of a block showing the index of available nitrogen and phosphorus by villages, based upon soil test summaries. (The numbers represent the villages.)

[1]Seth, Subbiah B.V., and Tamhane, R.V., *Journ. Ind. Soc. Soil Sci.*, 5, No. 2, 1957.

FIG. Soil test in progress with equipment carried in Mobile Soil Testing Laboratory.

standing for 15 minutes. Five drops of an indicator solution (made with phenol red, alizarin red-S and thymol blue) are added. The suspension is then shaken thoroughly and allowed to settle for 30 minutes. The colour is read against a calibrated colour chart.

2. *Available Phosphorus* is determined by taking 5 gm of powdered soil sample and shaking it with 10 cc of 0.5 M sodium bicarbonate solution and a pinch of Darco G-60 activated carbon. The suspension is filtered and a 2 cc aliquot of the solution is mixed with 2 cc of prepared ammonium molybdate solution and 1 cc of stannous chloride solution. The blue colour that is developed in the presence of phosphorus is matched against a calibrated permanent colour chart.

3. *Available Potash* is estimated by shaking five grams of powdered soil with 10 cc of a solution of acetic acid and sodium acetate, adjusted to pH 4.8, and filtering it. Two cc of the soil extract is made up to 10 cc by dilution with water. Two cc of a mixture (1 :1) of isopropyl alcohol and methyl alcohol are taken in a glass vial and six drops of cobaltinitrite solution are added. Two cc of the soil extract is then forced into the alcohol-cobaltinitrite solution with a hypodermic syringe. The turbidity that is developed is measured by matching with a calibrated chart.

REFERENCES

Spurway, C.H. and Lawton, K., "Soil Testing—A Practical System of Soil Fertility Diagnosis," *Tech. Bull. No.132*, Michigan State Coll. Expt. Stn., 1949.

Lunt, H.A. Swanson, C.L.W. and Jacobson, H.G.M., "The Morgan Soil Testing System," *Connecticut Agric. Expt. Stn. Bull No. 541*, New Haven, 1950.

Anonymous, "The Purdue Soil and Plant Tissue Tests," *Purdue Agric. Expt. Stn. Bull No. 584*, Lafeyette, Indiana, 1952.

CHAPTER XII

LOSSES IN THE SOIL BANK

SOIL CONSERVATION

The fertility of normal cultivated soil is largely associated with the top soil covering the land surface. It is no exaggeration to say that the world exists on the top nine inches of the earth's surface, because this statement is based on the fact that almost the entire production of vegetable matter on which animal and human life depends is cropped out of this fertile top soil. The fertility of the soil and, in addition, the very soil itself are subject to losses with cultivation and cropping. While the fertility is lost by the removal of plant nutrients through cropping and losses through leaching, the soil itself suffers from loss through erosion and its removal from the place of its formation by the action of wind and water. Even under normal conditions, where there is no disturbance of the soil surface through cultivation, there is a certain amount of erosion and loss by the action of wind and water, but this erosion is accelerated through cultivation and cropping, and can assume serious proportions under bad management of these operations. Geologists estimate that the earth's surface is being eroded by water normally at the rate of 1 inch every 200 to 500 years. Under conditions of accelerated erosion, as under cultivation, this rate of loss is increased manifold.

In various parts of the world, records exist of the damage caused by uncontrolled soil erosion. Such erosions have caused widespread havoc in many regions, like Africa, the Middle East, the U.S.A. and India. Where previously fertile tracts supporting prosperous agricultural communities existed, one finds now abandoned farms and agricultural lands converted into near deserts. The population and the human communities dependent upon those lands have had to abandon their homes and farmsteads and migrate to other places. Looking back into history, one finds that flourishing civilizations like those of Babylon, Egypt and Greece decayed and disappeared because the fertile valleys which provided food for the populace were chocked or silted up due to constant soil erosion. The decline of the productivity of the fertile lands and the loss of the crop-sustaining soil through erosion, for which no remedy had been worked out, hastened the disappearance of the population and with them their civilization. The menace of these losses, threatening the existence of every agricultural community, whether they are located in valleys, broad plains or the slopes of the hill ranges, has come to be realized in increasing measure in recent decades. The importance of maintenance and conservation of natural soil resources to safeguard the agricultural economy, through control of soil erosion to the maximum extent possible, has come to be realised increasingly in every progressive country.

CAUSES OF SOIL EROSION

Soil erosion is caused by the removal of soil particles from the location of their formation by the action of water and wind. The extent of the soil loss through erosion differs and depends upon a number of factors. When land is brought under cultivation, a larger amount of soil particles are exposed to the forces of erosion. The rate of movement of soil down a slope is greatly hastened and the amount of soil lost by erosion increases with the degree and length of slope and the rate and volume of flow of water, which in their turn are related to the intensity of rainfall. It also depends upon the rate of infiltration of water into the soil, and this rate depends upon the character of the soil. In a light textured soil, the infiltration rate is more than in a heavy textured soil. This rate is much less on a soil that has been under cultivation for a number of years than on a newly ploughed soil that had been under grass.

MECHANICS OF EROSION BY WATER

Most of the soil erosion in cultivated fields is caused by falling raindrops on the soil surface. This is referred to as "rain-splash." When a drop of rain falls on the ground, it first dislodges the particles of the soil and then it breaks down the soil aggregates by its momentum. As a result of the splash, the particles of soil are lifted into the air as much as 2 feet high and are transported horizontally 2 to 3 feet. The transported material is then carried away by the runoff. The extent of loss of valuable top soil on unprotected land has been estimated at more than 50 tons of top soil per acre on the average through the course of one year, and this may go up as high as 120 tons of fertile soil in a single year. The rate of this loss is low from a ground well covered by grass and is high on lands which are cultivated and cropped and varies with the type of cropping.

FIG. 77. Soil erosion causes serious loss of valuable top soil from cultivated lands.

TYPES OF EROSION BY WATER

There are four types of erosion concerning cultivated lands. These are: (i) sheet erosion, (ii) rill erosion, (iii) gully erosion and (iv) stream bank erosion.

SHEET EROSION

This type of erosion occurs on gently sloping lands. As the water flows over the land, the finer particles of the soil get mixed up with the running water and are carried away. This results in the gradual depletion of the top soil, year by year, until patches of subsoil appear in the field here and there. This type of insiduous erosion is taking place all the time and is noticeable by the washing away of top soil leaving thin furrows on the surface of the fields. If this is not checked in time, all the top soil may be washed out in course of time and the farmer may be only cultivating the infertile subsoil.

RILL EROSION

Rill erosion is an extension of sheet erosion. Where the land is steep or uneven, rills, i.e. small finger-like ditches make their appearance. First, they are small and do not interfere much with normal

FIG. 78. Rill erosion with its finger-like outlets through which fertile top soil is washed out of the field.

agricultural operations, but gradually they grow deeper and wider, thus hindering field operations and lessening the area for cultivation.

GULLY EROSION

If rill erosion is not controlled in time, the rills become wider and deeper, and form big gullies. Adjacent gullies, particularly near the drainage lines and the stream banks may then join up to form a big network of gullies and ultimately ruin an appreciable part of the agricultural area. The development of gullies, particularly of the deep type, is associated with certain soil characters. It is

FIG. 79. Valuable cultivable land has been lost through gully erosion.

a common sight in several states including Uttar Pradesh, Madhya Pradesh, Rajasthan and Gujarat to see extensive areas of cultivable land along the Jumna, Chambal, Sabarmati and Mahi rivers, affected by this type of erosion which has rendered large acreages unfit for cultivation. Relatively less extensive, but fairly intensive damage to agricultural lands is seen in gullied lands in Orissa, Mysore and Madras States. These large gullies or ravines encroach annually into the neighbouring cultivated lands and cause increasing damage. It is estimated that in these states between 40 and 50 lakh acres are thus rendered unfit and are lost to cultivation.

STREAM BANK EROSION

This is found on the banks of streams and rivers and on the agricultural lands close to them. The flowing river waters when they move with high velocity, gradually erode the banks causing collapse of the soil of adjoining agricultural lands. At times the rivers may flood over the banks, resulting in deposition of sand, and silt or in the cutting down of its banks. Every year, as a result of floods, various rivers erode large areas of land along their courses. The erosion caused along the banks of rivers like Kosi and Chambal, Damodar, Mayurakshi and Machkund and their tributaries, to name a few, leave behind many acres of agricultural land badly eroded and cut up by the forces of water.

FIG. 80. Stream bank erosion causes extensive soil loss. The adjoining agricultural lands get rapidly affected by the encroaching gullies.

EROSION BY WIND

When wind blows across land, it carries away particles of soil and deposits them in distant places. This will cause damage in two ways. First, the eroded field loses its fertile top soil, which, in course of time, can cause total infertility of the land, and secondly, the sand and coarse particles of soil carried in the wind stream may get deposited on good productive fields, thereby covering them and reducing or destroying their productive capacity. The dust storms, so common in the north-western parts of India, are instances of wind carrying away eroded soil particles for eventual deposition in other areas.

MECHANICS OF WIND EROSION

The mechanics of wind erosion have been studied closely in Canada and the U.S.A. The blowing wind has velocity as well as turbulence. It is the latter that is largely responsible for dislodging the soil particles from the ground and lifting them into eddies of air. Once in the air, the particles are made to move forward with the velocity of the wind. The velocity increases with height and this increase is called the "drag velocity."

The motion of particles by wind action may be of three types: (i) surface creep, (ii) saltation, and (iii) suspension.

Surface creep

This is confined to large particles which do not rise from the surface but move and roll on the surface of the ground. The particles affected in surface creep are the largest of the eroded particles and range from 0.5 to 1 mm in diameter. Usually, 5 to 25 per cent of the wind-eroded particles move by surface creep.

Saltation

In saltation movement, the particles are first dislodged from the soil surface, rise steeply into the air, move a short distance and bounce back on to the surface. They may then lose most of the energy by hitting other particles or continue in saltation by rebounding on the air. Most of the eroded particles move by saltation, nearly 50-75 per cent constituting this type of movement. The size of particles carried by saltation ranges from 0.05 to 0.5 mm in diameter.

Suspension

The fine dust, containing particles below 0.05 mm in diameter, is usually more resistent to erosion, but is affected by saltation movements of the larger particles. Particles in saltation stir up the fine

Fig. 81. Cultivation of land, and cropping up and down the slopes are sure methods to cause erosion and loss of soil and fertility.

dust and push them into the turbulent layers above where they are carried away into the air stream as fine dust. Due to the existence of a slow-moving viscous layer of air close to the surface of ground, the smaller dust particles which are usually aggregated, offer resistance to the action of wind. But once they are lifted into the air by the saltation movement of the bigger particles, the dust is taken to great heights and can only come down at great distances or through the action of rain. About 3 to 4 per cent of the eroded soil particles move by suspension.

FACTORS CONDUCIVE TO SOIL EROSION

Two main causes for accelerated soil erosion are: (1) the removal of natural vegetative cover and the consequent exposure of vulnerable top soil to the action of rain and wind, and (2) effect of improper tillage. The different factors that operate in the process of erosion of the top soil in a cultivated field may be classified as: (i) flow of water, (ii) topography and slope, (iii) effect of vegetation and cultural practices, (iv) nature of the soil, and (v) forces of wind. The influence and extent of each of these factors in soil erosion can be briefly considered here.

Water

This is the most common and important causative factor of erosion in agricultural land. The rain drops when they fall directly on the surface of the soil, loosen and transport the soil particles due to the splash effect. The intensity of the rainfall rather than the total amount of rainfall is the more important factor. The velocity of flowing water is also another important factor as its soil-carrying capacity is rapidly increased and multiplied. It is estimated that the erosive power increases 64 times, when the velocity is increased twofold, and when the velocity of flowing water is trebled, the soil-carrying capacity goes up to the almost unbelievable figure of 729 times.

Topography and slope

The velocity of flowing water increases as it moves down a slope. Hence, if the slope is a steep one or a long one, the chances of increase in the flow velocity are increased considerably, and accordingly, the erosion is also substantial. If the slope is uniform over a wide area, there will be only sheet erosion. If, on the other hand, the slopes are multilateral, then there will be both sheet erosion as well as gully erosion.

Effect of Vegetation and Cultural Practices

Undoubtedly, one of the chief causes of accelerated soil erosion is man's indiscriminate use of land. Forest and natural vegetation can protect the land surface, and when these protective covers are cut down or removed, the land is laid bare and exposed to the destructive action of the rain drops, which in torrential rain can tear up the surface into innumerable rills. The indiscriminate grazing of grass cover by cattle, is a major source of damage to the soil. Particularly, in arid regions where the grass cover is thin and the regeneration of vegetation is slow, overgrazing and the removal of the natural vegetation cover from the surface of the ground, are the causes of accelerated erosion. Indiscriminate cultivation and ploughing, combined with improper cropping practices carried out without consideration for protection

FIG. 82. Once a good forest land, indiscriminate cutting of the trees, clipping off of the branches of the remaining trees and uncontrolled grazing have laid bare the ground, resulting in extensive erosion. The subsoil is exposed and the land has come to be strewn with stones and gravel.

of the land, result in leaving the land exposed, with no protection against the natural erosive forces of wind and rain.

Nature of soil

The nature of the soil is also one of the most important factors in soil erosion. Soils with good permeability, and the power to absorb and drain rain water easily, are less prone to erosion compared to other soils which are less permeable. In general, the porous, light-textured red soils of peninsular India are not affected to such a great extent as the relatively impervious heavy black soils of the Deccan Plateau. This is specially so if the soils are deep as in the case of the deep and medium deep black soils of the Deccan area. Shallow soils underlain with an impermeable subsoil of a hard rocky or a heavy clayey nature are as susceptible to rapid erosion, as shallow soils with a highly permeable sandy subsoil.

Wind

Next to water, wind is the most important factor in soil erosion. The dry soil particles on the surface of soils are apt to be easily dislodged and blown by wind, unless sufficient protection is available. Wind erosion is most common in dry, open, level country, since the natural undulations of the ground surface—which reduce and, to an extent, prevent the action of the forces of wind—are absent.

Extent of Erosion

Critical studies to assess the degree and extent of soil erosion by the various forces of erosion over specified areas are being conducted in different countries since recent years only and, as such, the volume of information collected and made available is not large. However, such information, as is already available, indicates that the extent of soil losses from cultivated lands is much more than popularly believed. A sample survey conducted in the old Bombay State revealed that over 70 per cent of cultivated lands are affected by soil erosion and that 32 per cent are severely eroded. While the above information relates to the areas of black soils surveyed, it is easy to see that in other parts of the country, too, similar effects of erosion exist. The extensive erosion has undoubtedly led to the reduction in the useful areas available for cultivation, besides lessening the fertility of the top soil. It is estimated that nearly 50% of the total cultivated acreage in India (about 200 million acres) is affected by varying degrees of erosion.

Besides the natural forces of erosion, the methods of cultivation, cropping and crop rotations influence in a large measure the accentuation of the degree of erosion. A study made in Sholapur, on a medium deep black soil during the course of ten years (1931-1941) has indicated the following annual erosion losses under different systems of land use.

METHODS OF CONTROLLING EROSION IN CULTIVATED LANDS

For a proper control of erosion, one of the main points to be kept in mind is that the use of improved techniques in ploughing, land management, fertilization, crop rotation and irrigation are the best safeguards against soil erosion. A preliminary study of the soils of an area can give information about the land use capability, and furnish information about the type of cropping and rotation, and the erosion hazards that are present.

TABLE XXXV

ANNUAL LOSS OF SOIL UNDER DIFFERENT SYSTEMS OF LAND USES SHOLAPUR (1931-41)

Land Use	*Soil lost in tons per acre*	*Estimated time to erode all to soil (years)*
Under natural vegetation	0.53	1850
Trimmed	18	37
Kharif (bajra-tur)	24	42
Shallow cultivation	25	40
Deep cultivation	33	31
Rabi jowar	37	27

The methods aimed at controlling erosion of the soil have in view the conservation of the soil by preventing rapid run-off of rain water and reducing the flow of water on the surface. These practices serve another use in conserving the water also. These methods can be classified as follows: (1) provision of waterways, (2) small contour ploughing, (3) strip cropping, (4) contour bunding, (5) terracing, (6) provision of windbreaks, (7) cultural practices, and (8) use of soil conditioners—natural and artificial.

PROVISION OF WATERWAYS

Normally, in all cropped fields, and particularly those on sloping terrain, provision should exist for carrying away the surplus storm waters that drain from the land. The natural waterways which carry away the excess run-off from sloping fields should be kept in trim and workable condition. The size and shape of such waterways depend upon the local conditions but should be so regulated that they can carry away all the surplus water. There should be free discharge of running water, which should be estimated on the basis of high rainfall years. There should be no irregularities in the running course or the flowing surface to cause turbulence of water, because such turbulence

FIG. 83. Grassed waterways ensure that the water that is led out of bunds or terraces in fields does not cause erosion. Good grassed waterways ensure more complete absorption of the surplus run-off waters.

increases the scouring action of water. It is recommended that the waterways should have close vegetative covering like grasses or legume grass mixture. In order to establish this cover quickly, land preparation should be speedily got done while seeding and fertilization at double rates may have to be done on the waterways so that the grasses may be properly established before the rains commence.

CONTOUR PLOUGHING

This is one of the simplest methods of erosion control, but is effective only on land with 2 to 8 per cent slopes, on which it reduces erosion by about 50 per cent. The ploughing is done following the contours of the field so that the ridges formed provide sufficient resistance to the flow of rain water. Additional benefits may also be achieved, as contouring alone can increase the yields by 10 to 50 per cent. But this method of cultivation may not be effective on steeper slopes or on land which has a highly impermeable sub-soil. In such cases, in addition to contouring, other devices such as terracing may be necessary.

FIG. 84. Contour ploughing and cultivation safeguards this land against soil erosion by rain water. Rain falling on the land is made to move along the furrows and cannot run away fast carrying with it the soil and plant nutrients.

STRIP CROPPING

In this method, the fields are cropped in alternative strips. Usually, two types of crops are grown for the convenience of field operations. One crop will usually be an erosion permitting crop like jowar or maize which does not give enough ground cover. and another a close growing erosion resisting crop, like groundnut. There are three methods of strip cropping, viz. contour strip cropping, field strip cropping and buffer strip cropping. The contour strip cropping method is the most useful and works very well in well drained, cultivated fields where the slope is between 2 and 10 per cent. The width of the strips requires to be adjusted to suit local conditions, to provide for steepness of slope, nature of crop and prevailing rainfall duration and intensities. Usually, the sizes in common use are 100 feet wide strips for slopes less than 6 per cent, 80 feet for slopes between 6 and 10 per cent and 50 feet for steeper slopes. In Maharashtra strips about 72 ft. in width for erosion-permitting crops, alternating with strips 24 ft. in width for erosion-resisting crops are recommended.

The method of field strip cropping is only useful when the ground is too irregular for contour strip cropping. In this case, the crops are grown in strips across the slope of the land which may not necessarily be on a contour. The protection thus offered is limited and, therefore, this type of cropping should be combined with other improved cultural and cultivation methods where the terrain demands such processes.

In buffer strip cropping, protective strips are usually laid out in between contour strips to check and prevent erosion losses. The buffer strips are cropped with grass or legumes. This method is not as effective or as good as the first two, but may be used as a temporary measure.

CONTOUR BUNDING

In this method of conservation, bunds or terraces are put on the field in contours at suitable intervals to cover the entire field. Such bunds are specially helpful in low rainfall areas as this mehod provides for water conservation and also ensures that whatever rain falls on the fields is largely retained.

A contour bund performs three important functions, viz. (i) holds all available rainwater for a longer time in contact with the soil, thereby increasing the absorption of water into the ground; (ii) checks the velocity of runoff water; (iii) prevents erosion damage by the run-off water.

For a proper layout of contour bunds on a field, several factors have to be taken into consideration and these are: (i) type and depth of soil; (ii) slope of land; (iii) rainfall.

(1) *Type of soil*

Contour bunds have been found to be stable in red soils and also in shallow, medium and medium-deep black soils. In this type of depth categorization, as suited to soil conservation, soils with 9″ depth are considered shallow, those with 9 to 18″ depth as

FIG. 85. Crops raised in the contour in broad strips. Note also the bunds along the contour. Alteration of strips of crops with shallow and deep roots ensures better utilization of the soil moisture and plant nutrients besides ensuring soil conservation.

medium, and those with up to 36″ deep as medium deep soils. The bunds are not stable in heavy black soils, as they tend to develop cracks during the dry season, and these become sources of weak spots to cause breaches when the showers fall.

Slope

The bunds are spaced in accordance with the degree of slope. If the spacing is too narrow, farming operations may become difficult. The size and section of the bunds need to be modified to suit the degree of slope. The optimum spacing chosen should be such as to provide for an even distribution of the moisture collected on the land surface. The following relationship between slope and width of plots has been found to be satisfactory.

TABLE XXXVI

Slope per cent	*Vertical Interval in ft.*	*Approximate Horizontal Distance between Bunds*
0—1	3½	250 ft.
1—1½	4	320 ft.
1½—2	4½	250 ft.
2—3	5	200 ft.
3—4	5½	170 ft.
4—5	6	130 ft.

The size and section of bunds used on different areas are to be determined by taking into consideration the soil types and their depth. The section of the bunds recommended for red soils (sandy loam, loamy sands, loam) and black soils (silty clay, clay loam and clay) range between 10.0 and 16.0 sq. ft. for soils with depths ranging from shallow to medium deep. The bottom width of the bunds, which are trapez oidal in section, range from 6.0 to 10.0 feet with a top width of 1.5 feet and height between

FIG. 86. Contour bunding. The alignment is formed and soil is shifted to form the bund which will slow down the flow of rain-water which will otherwise rush down the slope.

2.5 and 3.0 feet, the side slopes having ratios between 1 : 1 and 1.5 :1. The material for bunds is dug out of burrow pits, which may range between 10 and 15 feet in width all along the line of the projected bund and removed about 10 feet from it. The burrow pits may be located upstream in the case of shallow and medium soils or down stream if the soils are deep. The cost of construction of such bunds varies from Rs. 30 to Rs. 70 per acre depending upon the nature of soil and the spacing of the bunds. These bunds require to be maintained and protected against grazing animals, passage of carts, etc. It is possible to grow on the bunds *tur* or castor or soil protection creepers.

Rainfall

The nature and intensity of rainfall conditions will influence greatly the type of bund to be put up. It will be necessary to have different types of bund to suit low rainfall tracts and tracts with high rainfall. Narrow and small cross section bunds may be adequate in scarcity areas, while heavier section bunds may be needed in areas of higher rainfall. Also, surplussing management, like providing outlets and weirs may be necessary where the rainfall is more than 15 inches.

TERRACING

Terracing is a permanent and satisfactory means of soil conservation though it is more expensive. In this method the field is formed into descending flat terraces of some suitable width depending upon the slope. This method of soil conservation has been in vogue in India in one form or another from ancient times, specially where paddy is grown on sloping ground. Terraces are of two types, viz. (i) absorption type; (ii) drainage type.

In the absorption type, greater emphasis is placed on the conservation of run-off by putting ridges on contours across the line of slope. The flowing water is held back and time is given for the absorption of moisture. In the United States, these terraces are commonly used, and are referred to as "broad based" bunds. The broad base admits of machinery being moved over it, and is, therefore, convenient for adoption in countries where mechanised cultivation is in vogue. The broad-based bunds are also tilled and crops raised, so that land space occupied by the terraces is not wasted. These terraces, however, require to be kept

FIG. 87. Bund formed in the field along the contour by borrowing soil from both sides of the bund. Such bunds ensure that erosion by rapid run-off of the rain water is minimised. They help better absorption of the rain-water received on the fields. The bunds require to be maintained and protected by raising grass and crops like castor which can also serve as wind-breaks.

in shape during operations in successive cropping seasons. In India we have "narrow based terraces," locally called "bunds," which do not permit of cultivation, and to that extent, area of the cultivable land is reduced.

In the drainage type, the excess of runoff is led through a gently sloping channel, while in the absorption type, the ridge is of primary importance. In the latter type, the channel is important.

Several factors determine the size and construction of terraces. Usually, the width of the terraces decreases as the steepness of the slope increases. The vertical interval between two terraces may be arrived at by the formula:

$$V.I. = \frac{S}{3} + 2$$

When V.I. = vertical interval, S=percentage of slope. While

the above formula is used for heavy-textured soil areas, a slightly different formula V.I. $= \frac{S}{2} + 2$ is used in areas of light-textured soils. The vertical interval may be converted into horizontal distance by multiplying it by $\frac{100}{S}$. The length of a single terrace should not normally exceed 1600 feet and if this is not

FIG. 88. Terraces formed on steep sloping land. Such terraces ensure that the soil loss is kept at a minimum and the cultivators are enabled to raise crops successfully. Without such protection the lands on steep slopes should not be permitted to be cultivated. They are otherwise best kept under forests or permanent vegetation.

possible, later on, cross terraces may be introduced at intervals to prevent movement of water from one end to the other. The terraces can be made either mechanically with the help of a bull-dozer or tractor with disc tiller attachments with bullock power. The broad-based terrace admits of crop cultivation but maintenance under cultivation by non-mechanical means is somewhat difficult. Such bunds or terraces show a possibility of being adopted in deep black soil areas, in suitable locations, the terrace portion being put under grass. In India, however, the narrow based bund formed mainly by manual labour is the most

FIG. 89. Improved terracing methods practised in Nilgiri District on lands with slopes up to 33%. Note the inward slope in the terrace bed. The terrace faces and the bunds are grown over by grass. A good crop of potato is seen in the foreground.

common practice. It is very important that proper attention is given to the maintenance of the bunds and not to allow formation of silt bars in the channels or scouring in ridges.

In terrains with slopes in excess of 15-25 per cent, different types of terraces are needed. Stepwise terraces separated at suitable interfalls are formed. Such constructions are referred to as "bench terraces" and consist of a series of platforms along contours separated by vertical drops or risers. Bench terracing work depends upon the suitability of the land, percentage slope, type, depth of soil and land for farming practices. Bench terraces may be level, or slope inwards or outwards, depending upon the total rainfall. Formation of bench terraces is enforced in the Nilgiri District of Madras State where intensive cultivation of lands with slope in excess of 25% is in vogue, and cultivation of crops on such terraces is permissible up to 33% slope. Considerable work has been done to work out the economic spacing, grade, length and cross-section of the terraces to be farmed on such terrain, and the tentative formula arrived at the Soil Conservation Research Station at Ootacamund is as follows:

$$\text{V.I.} = \frac{\text{WS}}{(100-\text{S})},$$

When V.I. is the vertical interval, W the width of terrace and S the slope percentage. The cost of bench terracing under the Nilgiris conditions ranges between Rs. 400/- and 800/- per acre. In spite of the higher initial cost involved, the experiences of farmers show that the increased outturn of a cash crop like potato, that is grown there, enables this expenditure to be recovered in 2 to 6 cropping seasons.

PROVISION OF WIND-BREAKS AND SHELTER-BREAKS

Wind-breaks and shelter-breaks are provided to act as barriers against the forces of wind. A shelter belt is stronger and more extensive than a wind-break which is merely a line of shrubs grown across the line of wind. A shelter-belt consists of a combination of trees, shrubs and closely growing vegetation to check the force of wind, to protect against aerial and animal depredations and, to some extent, conserve soil moisture. A good shelter-belt consists of 2 to 3 rows, the first row consisting of quick-growing shrubs, the

Fig. 90. Line of tall trees growing along the edges of the field act as windbreaks and protect the soil against wind erosion.

second row consisting of quick-growing trees and the third row of slow-growing but long standing trees.

CULTURAL PRACTICES

Erosion in cultivated lands can be controlled through the regulation of cropping patterns, crop rotations and other cultivation practices that are adopted on the land. These may be described under the following three measures:

1. Rotations
2. Stubble mulch cultivation
3. Irrigation methods

Rotations

The cultivation of land by successive reapings of a crop, which leaves the land exposed to erosion, can accelerate the erosion loss. Repeated cultivation with soil-depleting or erosion-permitting crops like jowar, ragi, maize, and cotton can easily cause not only the rapid loss of fertility but also of the top soil. Soil-binding and fertility-building closely grown crops, like groundnut, and leguminous crops, like pulses can offset these losses and a proper rotation or combination of the two types of crops should be grown to minimise the soil losses.

Experiments conducted in the United States have shown that cropping continuously with row crops results in soil losses which are four to five times as much as when a spring grain and hay are rotated with row crops.

Stubble mulch cultivation

Farmers in many parts of the world are accustomed to clean cultivation. In stubble mulch farming, a mulch of crop residue

FIG. 91. Good growth of grass and vegetation has protected the lands against further encroachment and loss of land through stream bank erosion.

litter is left on the surface of the ground. Leaving crop residues such as jowar and cotton stalks on the soil instead of their complete removal will reduce soil losses to as much as 50 per cent. This will also reduce the erosion hazards by wind. The advantages claimed for this method include protection against soil loss, better absorption of rainfall, reduced soil evaporation loss and increased yields.

Irrigation method

Irrigation on slopes is apt to result in soil erosion if conducted in an uncontrolled fashion. The row crops and plants or trees, that are to be irrigated should preferably be on the contours so that the speed of water-flow is restricted and controlled so as to avoid erosion. Various methods of irrigation are available to control and reduce erosion such as the border method, check method and flooding, each depending upon special circumstances such as steepness of slope, soil type, nature of crop grown, etc. A proper choice of the method can permit a rational use of the irrigation facilities.

SOIL CONDITIONERS—NATURAL AND ARTIFICIAL

A proper condition of the surface of the soil with an adequate aggregation of the particles of top soil and development of suitable structure in the subsoil can ensure that the water that falls on the surface soaks in rapidly and is drained off. Such a set of conditions can ensure that erosion by water-flow is minimal. The aim of adding organic matter through manuring, rotation of crops and the adoption of improved cultivation practices, are aimed at obtaining these conditions in a natural manner. Attempts have been made to utilize various chemicals to produce suitable aggregates in the soil particles and also to build up a satisfactory structure in them. Treatment with such chemicals which are based on polyuronide or polyacrylonitrile compositions has been suggested as artificial soil conditioners which, besides having other advantages, can afford protection to the soil against erosion. However, in view of the heavy cost of the chemicals involved when field scale applications are considered, the methods have not proved economical. Further, the protection afforded is also of limited duration and so the success in the application of synthetic soil conditioners to prevent erosion is doubtful.

The use of bitumen and compositions based on this material have been considered for stabilizing sandy soils, as in sand dunes and sandy desert areas. Bitumen emulsions have been used to spray the surface of shifting sands, so that the particles are aggregated or cemented together by the solidifying bitumen and thereby protected against the devastating activity of winds. The possibility of establishing grasses and trees after thus stabilizing the sandy surface has been under investigation in the sand dune affected areas in south-eastern parts of the peninsula and in the desert regions of

FIG. 92. Soil erosion along the river bank. The soil removed through such rivers can vitally affect the life of reservoirs formed in their course for power generation or irrigation. Growth of protective vegetation cover on the banks can protect against serious soil loss. Easing the slopes, forming terraces and cultivation of crops like paddy, where rainfall conditions permit such cropping, can ensure better protection to such gradually sloping banks.

Rajasthan. The material used in these treatments is relatively inexpensive and some measure of success is claimed for stabilizing the sand dune surface with such bitumen emulsion treatments.

REFERENCES

STALLINGS, J.H., *Soil Conservation*, Prentice Hall, New York, 1959.

GADKARY, D.A., *A Manual on Soil Conservation*, Deptt. of Agri., Poona, 1956.

RAMA RAO, M.S.V., "Soil Conservation in India," I.C.A.R., New Delhi, 1962.

GUSTAFSON, A.F., *Using and managing Soils*, McGraw Hill Book Co. 1948.

GUSTAFSON, A.F., *Conserving of the Soil*, McGraw Hill Book Co., 1937.

ARCHER, S.G., *Soil Conservation*, University of Oklahoma Press, 1956.

CHAPTER XIII

SOIL AND WATER USE—PROBLEMS OF IRRIGATED AGRICULTURE

WATER RESOURCES AND DEVELOPMENT OF CIVILISATIONS

The story of man's progress towards civilisation is intimately connected with his efforts to master and utilise the fresh water resources of the world. The rise of great civilisations like those of Mohenjodaro, Assyria, Babylonia, Greece, Rome and Egypt and of the flourishing agriculture they once sustained were due to the elaborate network of irrigation systems that they built up and to the careful management of their fresh water resources. Evidence of these structures is still to be seen in many parts of the world where these civilisations thrived, and some of these are in use even today. When, due to neglect and decadence, the care and maintenance of those systems were ignored, the productivity of their lands declined, leading to the steady disappearance of the civilisations. The silting up of the canal systems, the waterlogging of the irrigated lands and the development of salinity and alkalinity due to injudicious management were sure causes leading to the decay of the civilisations and their final disappearance.

Water sustains life. It is the mainstay of agriculture. But, unfortunately, the water resources of the world are not equally distributed on the surface and the sub-surface areas of our globe. Where there is less of it, great effort is needed to locate it, conserve it, and use it to the best possible advantage. Where there is more of it or an undesirable excess of it, it is necessary to get rid of the excess and prevent it from damaging the countryside and valuable agricultural land. And where there is none, it is to be conveyed from great distances at a high cost. A close study of the rainfall and groundwater potential of any region is necessary, before a successful farming programme suitable to it can be developed. This is the *Hydrologic* study of the region.

THE HYDROLOGICAL CYCLE

The hydrologic study of a region includes observations on atmospheric precipitations like rainfall, dew and snow, water flow from rivers and streams passing through the area, evaporation from land and water surfaces, transpiration from vegetation in the area, moisture content of surface and sub-surface soil.

Water that is received on the soil surface in the form of rain is wholly or partly absorbed by the soil and the process is known as *Infiltration*. The capacity to absorb and hold the water that is received varies from soil to soil and is known as the *Infiltration Rate*. Where rainfall is more than the infiltration rate, the excess becomes the

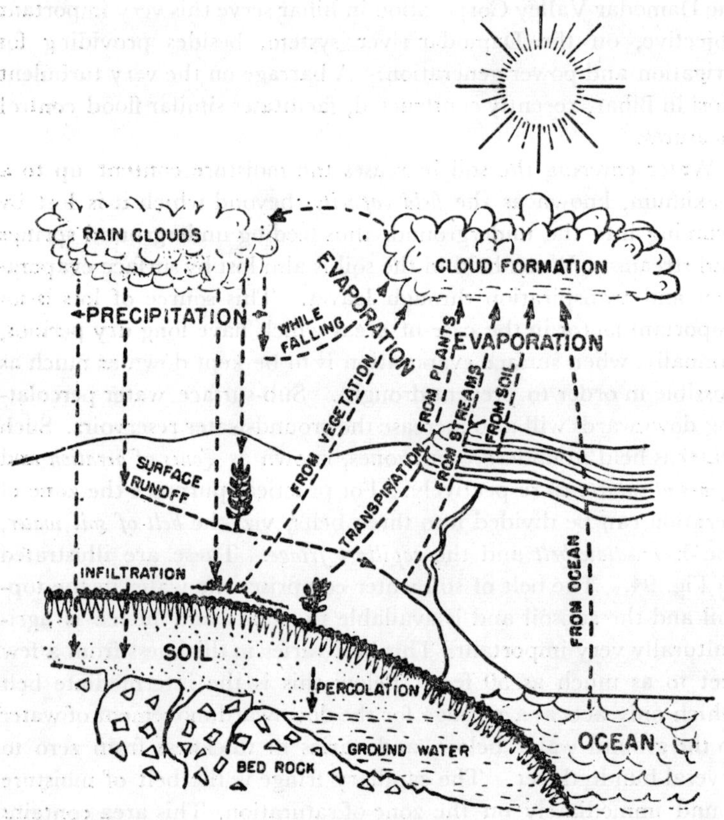

FIG. 93. The water cycle in nature

surface run-off which moves out into the natural water courses, streams and rivers. Excessive surface overflow is undesirable as it is likely to cause soil erosion. It is generally controlled by provision of suitable bunds and check-dams across the direction of flow and the excess thus stored in reservoirs is released gradually which can be used later on for irrigation, power generation and such like uses. In South India, such reservoirs of water known as *tanks* have been in existence for several centuries and are used principally for irrigation. The construction of tanks and reservoirs serve also as means of controlling floods. Large-scale flood control measures have in the recent past been taken up in different parts of the country covering many river systems. A series of dams put up by the Damodar Valley Corporation in Bihar serve this very important objective, on the Damodar river system, besides providing for irrigation and power generation. A barrage on the very turbulent Kosi in Bihar, recently constructed, facilitates similar flood control measures.

Water entering the soil increases the moisture content up to a maximum, known as the *field capacity*, beyond which it is lost by draining into the underground, thus feeding underground springs and streams. Water held in the soil is also lost by surface evaporation and transpiration through leaves. This source of loss is an important factor in the case of areas which have long dry periods, annually, when surface evaporation is to be kept down as much as possible in order to prevent drought. Sub-surface water percolating downwards will also increase the ground-water reservoirs. Such water is held in underground zones, known as *Zones of aeration* and *Zones of saturation* respectively. For practical purposes the zone of aeration can be divided into three belts, viz. the *belt of soil water*, the *intermediate belt* and the *capillary fringe*. These are illustrated in Fig. 94. The belt of soil water comprises the water in the topsoil and the subsoil and is available to plants and hence is agriculturally very important. This belt varies in thickness from a few feet to as much as 50 feet. Below this is the intermediate belt which only acts as a passage for the downward movement of water to the ground water below, and varies in thickness from zero to several hundred feet. The capillary fringe is the belt of moisture found immediately on the zone of saturation. This area contains water held by capillary forces. Its thickness varies from a few

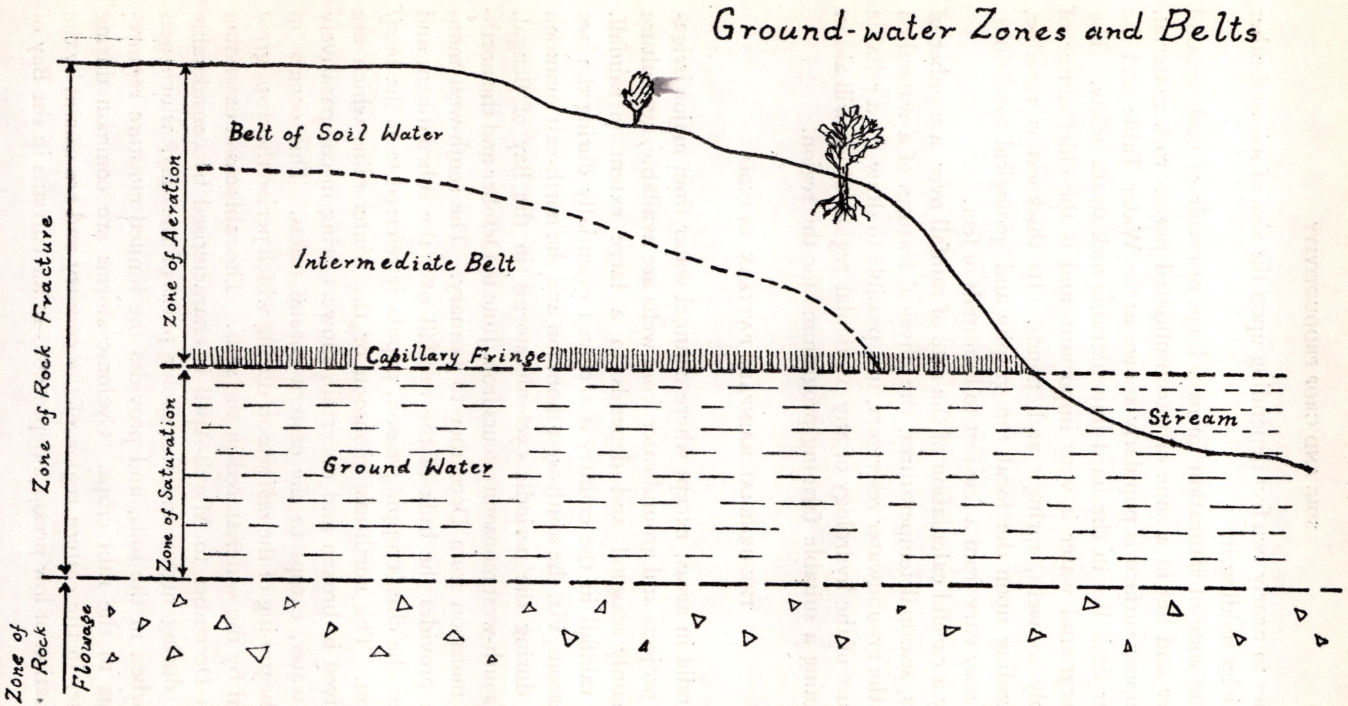

FIG. 94. Sketch showing the zones of ground water accumulation and the belts of soil water which may be available to plants.

inches to nearly two feet depending upon the size of soil and other particles holding it.

The zone of saturation forms a huge reservoir of underground water and lies in a zone of unconsolidated porous rock material. Its upper surface is popularly known as the Water Table and the lower limit lies in the hard impenetrable rock strata below. The underground water is very important and is the chief source of supply to wells, springs and ponds. Its thickness is variable, depending upon the local topography and geological conditions, and may vary from a few feet to hundreds of feet.

By a careful calculation of the rate of rainfall over a number of years, seasonal temperatures, the physical features of a watershed and the groundwater reservoirs, it is possible to draw up a reliable picture of the hydrology of any particular region which will aid in planning a suitable farming programme for the region.

THE GENERAL RAINFALL PATTERN IN INDIA

In India in areas, except where channel water from major irrigation projects and groundwater from wells are available, agriculture is mainly seasonal and depends to a large extent on rainfall. The rainfall in the country is received essentially during the two monsoons, viz. the south-west monsoon and the north-east monsoon and during the periodic cyclonic storms in the Bay of Bengal. The south-west monsoon occurs from June to October and the north-east monsoon from December to January. The south-west monsoon provides the bulk of the rainfall over the subcontinent and covers the chief cropping season, popularly referred to as the *kharif* season. The north-east monsoon or the winter rains which are received in broken and uncertain showers bring in comparatively less water, except in the eastern coastal tracts. This occurs in the beginning of the *rabi* season during which period the crops grow aided by the water stored in the soils. The rabi season extending from December to March-April is characterised by considerable mist during nights and the cooler parts of the day, which gets absorbed in the soils, and provides the limited moisture requirements to the rabi crops. Cyclonic storms are common during winter in the eastern region of the country and are caused by the formation of low pressure depressions in air currents in the Bay of

Bengal. The depressions usually tend to move westwards towards the east coast of peninsular India bringing sudden heavy downpour and causing considerable havoc to standing crops. The Western Coastal belt and the adjoining hilly areas of the peninsular and the submontane regions of the Himalayas receive the heaviest rainfall ranging from 60″ to over 150″. Rainfall as high as 460″ is met with in some areas as in Cherrapunji in Assam and over 350″ in Agumbe in Mysore. Rainfall decreases inland and becomes scanty and uncertain as in some of the dry districts of Rajasthan, Gujarat, Madhya Pradesh, Andhra Pradesh and Mysore.

CROPPING PATTERNS SUITED TO WATER AVAILABILITY

The pattern of rainfall being variable and the topography of the different cultivated areas also being highly variable, different systems of water utilisation and cropping practices which are adjusted to these varying conditions are in vogue. The main systems of water utilisation for crop production in India can be grouped as follows:

1. Perennial irrigation from major and minor irrigation projects—crops like paddy, sugarcane, cotton, wheat, jute and plantains are grown.
2. Irrigation from tanks which are fed seasonally from rains—crops like paddy, sugarcane, ragi, cocoanut, areca, and plantains.
3. Irrigation from wells—crops like vegetables, fruit crops, tobacco, wheat, paddy, ragi, and potatoes.
4. Occasional or protective irrigation for crops like jowar, cotton, pulses, oilseeds and millets.
5. Assured rainfall areas—crops grown include paddy, sugarcane, plantains, jute, coffee, rubber, tea, spices like cardamom, pepper, wheat, oilseeds, jowar, and mesta.
6. Uncertain rainfall and irrigation facilities not available—crops grown are millets, pulses, gram and oilseeds.

IRRIGATION TO SUPPLEMENT RAINFALL

Groundwater resources which are partly related to rainfall conditions determine to a large extent the intensity of irrigation in the case of well irrigation. The development of irrigation by tanks is linked with the topographic situation of the concerned areas. In areas of undulating terrain, it is possible to have a large number of tanks of varying sizes, while generally, in flat areas, as in large parts of Maharashtra and Madhya Pradesh, the development of irrigation facilities from such tanks is somewhat restricted. The areas coming under irrigation of different types in the various States of India are given in Table XXXVII.

TABLE XXXVII

NET AREA IRRIGATED IN INDIA—AREA IRRIGATED FROM DIFFERENT SOURCES

(*in thousand acres 1958-59*)

State	*Canals*		*Tanks*	*Wells*	*Other*	*Total*
	Govt.	*Private*			*sources*	*(net)*
Andhra Pradesh	3,087	22	2944	728	265	7046
Assam (excluding N.E.F.A.)	178	72	—	—	634	1533
Bihar	1,071	262	1024	681	1951	2090
Maharashtra & Gujarat	648	74	285	2484	116	3807
Jammu & Kashmir	143	560	—	7	26	736
Kerala	377	73	79	35	315	879
Madhya Pradesh	854	6	281	741	90	2072
Madras	2072	3	2076	1346	89	5586
Mysore	460	13	859	394	286	1922
Orissa	487	69	1223	94	541	2414
Punjab	4858	123	17	2322	41	361
Rajasthan	818	—	774	1944	35	3571
Uttar Pradesh	4583	28	1084	5754	688	12137
West Bengal	966	956	910	39	468	3339
Delhi	32	—	5	42	—	79
Himachal Pradesh	—	—	—	—	96	96
Manipur	—	149	—	—	—	149
Tripura	—	—	—	—	8	8

Source: Agricultural Situation in India, Vol. XVII, No. 8.

It is seen from the above table that considerable areas in Bombay, Madhya Pradesh, Punjab, Rajasthan and Uttar Pradesh depend largely upon groundwater sources for irrigation purposes.

INTRODUCTION OF IRRIGATION IN NEW AREAS

The investigations of any new project for irrigation require that an irrigation scheme proves useful, economical and successful. When such projects are considered in a new region, besides the engineering aspects of construction of dams or reservoirs to store the waters of rivers, and canals to lead waters thus stored to the cultivable lands, certain factors and broad aspects of the region require to be considered. Climate and rainfall, for example, are important

FIG. 95. Land-levelling is an important feature in introduction of irrigation. Mechanical means using scrapers can speed up the work of bringing the land surface to required level.

factors to be taken into account, since an arid tract with a low rainfall raises problems of high water losses, and hence the irrigation requirements will be higher compared to a less arid or a humid region. Similarly, the physical features of the irrigated tract also, to a large extent, control the economics of a project, which may weigh heavily in proving whether the project could be a success or not. If topography is highly undulating or rugged, it may be necessary to provide for costly embankments to carry the canals or provide deep cuttings and tunnels for the waters to pass through, besides providing elaborate drainage systems to prevent waterlogging of the valley bottoms and lowlying areas. Irrigation under such conditions may become costly and the benefits accruing therefrom may not be commensurate with the costs involved, or in other words attractive from an investment point of view.

IRRIGATION AND SOIL PROBLEMS

An important aspect of bringing new land under irrigation is the evaluation of the different qualities of soils and then, judging the value of irrigation on the land. The geological nature of the tract, and the various constituents present in the soil, besides the physical characteristics of the soils themselves, influence this value. The geological nature of the tract may be such that the minerals or salts present in the rock or the substratum may tend to move up to the soil after introduction of irrigation and tend to lower the agricultural qualities of the soils. Inherent physical factors of the soil, like water-holding capacity, permeability, the degree of sodium saturation in the exchange complex, and soluble salts present in the soil, may influence adversely the reaction of the land to irrigation. The lack of knowledge about these properties of soils, prior to irrigation, has in a number of cases resulted in the development of acute saline and alkaline conditions besides creating waterlogging and problems of health. In the black soil tract in the Nira canal area of Bombay State, where irrigation was first introduced towards the beginning of this century, the lands were severely damaged due to the development of salinity and alkalinity resulting from the inherent characteristics of the soils. This led to widespread distress among the agriculturists, many of whom faced ruin. These acute pro-

FIG. 96. Land-forming is essential in any programme of irrigated agriculture. A simple device using bullock power is in operation.

blems have come to be gradually remedied through the proper understanding, of the management problems involved. The lessons learned from such experiences have underlined the great need for scientific assessment of the properties of soils through surveys, prior to introduction of irrigation.

ASSESSING UTILITY OF IRRIGATION—PRE-IRRIGATION SURVEYS

When irrigation is to be introduced for the first time on land, which previously had remained uncultivated or cultivated only for rain-fed crops, it is necessary to assess first the physical and chemical characteristics of the soils and their potentialities under intensive cropping which will naturally follow the provision of irrigation facilities. This is done by carrying out a pre-irrigation survey. Such a survey gives information about soil characteristics, its depth, texture, structure of different horizons, intensity and extent of soil erosion, data on the existence of saline or alkaline conditions, drainage characteristics, depth of water table, nature of parent rock, topography and the direction of flow in natural waterways, occurrence of hard pan in the soil profile and similar pertinent information. Such information, furnished along with a soil map of the area, will be of great help in planning land use when irrigation eventually becomes available in the area.

In many cases, the introduction of irrigation through canal systems connected with major irrigation projects, as already indicated, brings into existence, a variety of problems. The development of salinity and waterlogging are some of the important problems to be faced with introduction of irrigation through such large-scale irrigation systems. The conduct of pre-irrigation surveys furnishes information about the potential dangers in this regard and the information can be used for taking adequate remedial measures through regulation of water supplies, choice of crops and cropping patterns or through means of reclamation, provision of drainage facilities and suchlike management measures.

GROUPING FOR IRRIGATION SUITABILITY

The information collected from pre-irrigation surveys is also used to classify the soils into different groups of capability or irrigation suitability. In view of the variety of factors involved in assessing the properties of the soil, in respect of its suitability for irrigation, the methods of evaluation and the characteristics taken

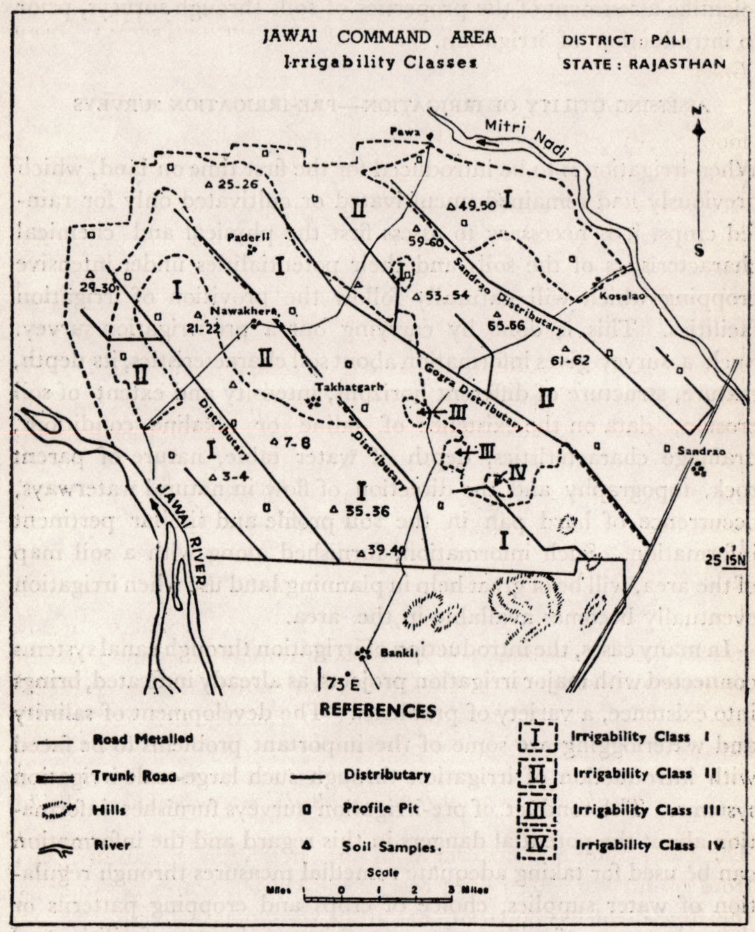

Fig. 97. Map of an area proposed for irrigation covered by soil survey as the lands are classified according to different irrigability classes.

into account in these procedures have varied a great deal from area to area. The standards adopted in different States in India have also varied a great deal, so that it is difficult to find common features among many of them. However, brief mention should be made here in regard to some of the standards that have been suggested for grouping lands according to their suitability for irrigation. The groupings suggested in the Soil Survey Manual (1960) of the Central Soil Conservation Board, India, include the following three groups.

Group I

Soils having good moisture-holding capacity (more than 1½" per foot of soil), low water table, soluble salts or exchangeable sodium present in negligible amounts, are quite suitable for irrigation. Absence of pan formation in the horizon, negligible salinity in subsoil water with good internal permeability, are favourable factors.

Group II

These soils are suitable for controlled irrigation, and have available moisture of medium capacity (between ¾ inch and 1½ inch per foot of soil), water table close to the surface, soluble salts and exchangeable sodium in moderate amounts. Internal permeability is fair and with no pan formation within the root zone. The subsoil water may be slightly to moderately saline.

Group III

Soils of this group are unsuitable for irrigation owing to the low depth of soil, presence of rocky strata or impervious pan formation at shallow depth, high pH and high salt content. Low available moisture capacity and low internal permeability also render such soils unsuitable for irrigation purposes.

Classification according to US Bureau of Reclamation

The particulars and criteria as to the irrigation suitability vary from country to country. In case of the United States of America, the Bureau of Reclamation has drawn up detailed standards for irrigational classification which have been applied in the various projects investigated by them. Their classification divides the soils into six classes, and of these, four are in respect of soils which are fit for irrigation, while the remaining two are not. Consideration of a complex of factors is made in these classifications, out of which the

[1]U.S. Dept. of Interior, *Bureau of Reclamation Manual*, 1953.

crop productivity of the land is given a place of importance. The characteristics of these six classes are given here.

Class I—Arable lands

Included in this class are lands that are highly suitable for irrigation farming, being capable of producing sustained and relatively high yields of a wide range of climatically adapted crops at reasonable cost. They possess smooth features and lie on gentle slopes. The soils are deep and of medium to fairly fine texture with mellow, open structure, allowing easy penetration of roots, air and water, and having free drainage, yet with good available moisture capacity. These soils are free from harmful accumulations of soluble salts and can be readily reclaimed. Both soil and topographic conditions are such that no specific farm drainage requirements are anticipated. Minimum erosion will result from irrigation, and land development can be accomplished at a relatively low cost. These lands potentially have a relatively high payment capacity.

Class 2—Arable lands

This class comprises lands of moderate suitability for irrigation farming, being measurably lower than those in Class I in productive capacity, adapted to somewhat narrower range of crops, more expensive to prepare for irrigation or more costly to farm. They may have a lower available moisture capacity, as indicated by coarse texture or limited soil depth, they may be slowly permeable to water because of clay layers of compaction on the subsoil, or they may be moderately saline which may limit productivity or involve moderate costs for leaching. Topographic limitations include uneven surface requiring moderate costs for levelling, small slopes requiring shorter length of runs, or steeper slopes necessitating special care and greater costs to irrigate and prevent erosion. Farm drainage may be required at a moderate cost, or loose rock or woody vegetation may have to be removed from the surface. Any one of the limitations may be sufficient to reduce the lands from Class I to Class 2 but frequently a combination of two or more of them is operating. Class 2 lands have an intermediate payment capacity.

Class 3—Arable lands

This class represents lands that are suitable for irrigation development, but are approaching marginality for irrigation, and are of distinctly restricted suitability because of more extreme deficiencies in the soil. These lands may have good topography, but, because of inferior soils they have restricted crop adaptability and require larger amounts of irrigation water or special irrigation practices, and demand greater fertilisation or more intensive soil improvement practices. They may have uneven topography, moderate to high concentration of salts or restricted drainage, susceptible to correction, but only at relatively high costs. Generally, greater risk may be involved in farming Class 3 lands, but, under proper management, they can have an adequate payment capacity.

Class 4—Limited Arable or Special Use

Lands are included in this class after special economic and engineering studies have shown them to be arable. They may have an excessive, specific deficiency or deficiencies susceptible to correction at high cost, but are suitable for irrigation because of existing or contemplated intensive cropping such as for truck and fruits; or they may have one or more excessive, non-correctable deficiencies thereby limiting their utility to meadow, pasture, orchard, or other relatively permanent crops. They are, however, capable of supporting a farm family and meeting water charges if operated in units of adequate size or in association with better lands. The deficiency may be inadequate drainage, excessive position (allowing periodic flooding or making water distribution and removal very difficult), rough topography, excessive quantities of loose rock on the surface or in the plow zone or vegetation cover such as timber. The magnitude of the correctable deficiency is worthy of requiring outlays of capital for land development in excess of those permissible for Class 3 lands because of the anticipated utility. The Class 4 lands may have a range of payment capacity greater than that of the associated arable lands.

Class 5—Non-arable

Lands in this class are non-arable under existing conditions, but have potential value sufficient to warrant tentative segregation for special study prior to completion of the classification or they are lands in existing projects whose arability is dependent upon additional scheduled project construction or land improvements.

They may have a specific soil deficiency, such as excessive salinity, very uneven topography, inadequate drainage, or excessive rock or tree cover. In the first instance, the deficiency or deficiencies of the land are of such nature and magnitude that special agronomic, economic or engineering studies are required to provide adequate information, such as extent and location of farm and project drains, or probable payment capacity under the anticipated land use, in order to complete the classification of the lands. The designation of Class 5 is tentative and must be changed to the proper arable class or Class 6 prior to completion of the land classification. In all instances, Class 5 lands are segregated only when the conditions existing in the area require consideration of such lands for a competent appraisal of the project possibilities, such as when an abundant supply of water or shortage of better lands exists, or when problems related to land development, rehabilitation and resettlement are involved.

Class 6—Non-arable

Lands in this class include those considered non-arable under the existing project because of their failure to meet the minimum requirements for the other classes of land, arable areas definitely not susceptible to delivery of irrigation water or to provision of project drainage, and Classes 4 and 5 lands when the extent of such lands or the detail of the particular investigation does not warrant their segregation. Generally, Class 6 comprises steep, rough, broken or badly eroded lands; lands with soils of very coarse or fine texture, or shallow soils over gravel, shale, sandstone, or hard pan, and lands that have adequate drainage and high concentrations of soluble salts or sodium. The Class 6 lands do not have a sufficient payment capacity to warrant consideration for irrigation.

EVALUATION OF SUITABILITY OF LAND FOR IRRIGATION

While classificational aspects to group soils according to their irrigation suitability have been followed in various countries, attempts have been made by numerous workers to evaluate mathematically these classificational groups by allotting values to various factors which influence the suitability of soils for irrigation. Different methods have been suggested for such evaluation, and a system that has been found useful in many areas is the one suggested by R.E. Storie and is known as the Storie Index.[1] The basis for assessing the agricultural value of soils by this method is to allot values to different grades of each of the soil and related factors

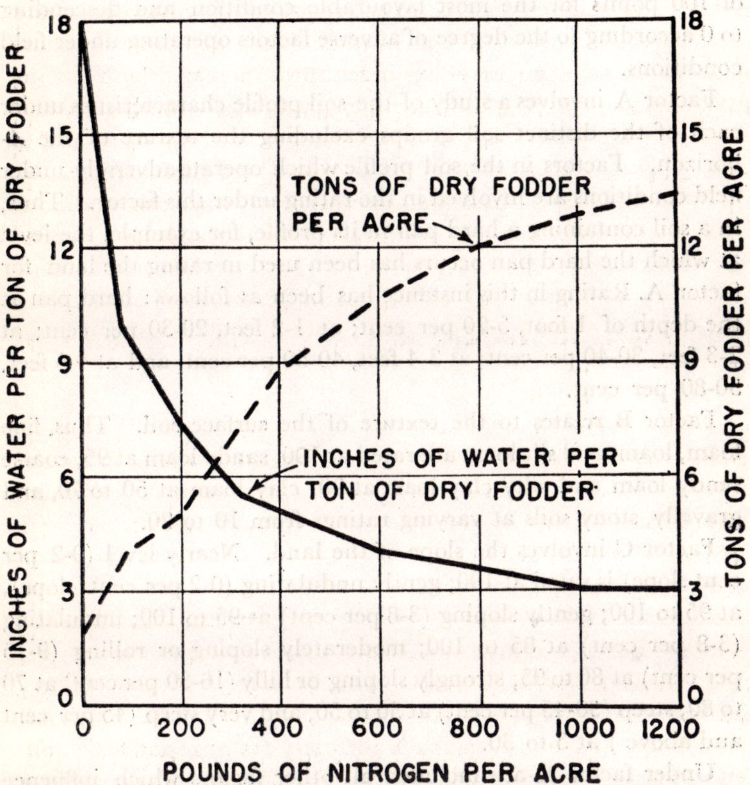

FIG. 98. With no nitrogen fertiliser, it required 18 acre-inches of water to produce one ton of dry Bermuda -grass fodder; whereas with 1,000 pounds of N per acre, only 3 acre-inches of water was required. (Phosphorus and potassium were adequate in the soil).

considered important in cultivation and cropping, and to sum them up using certain factors. The information with regard to the soil properties which are collected during a pre-irrigation survey, are

[1]Storie R. E., "An Index for Rating the Agricultural Value of Soils," *Calif. Agri. Stn. Bull., No. 556*, Berkeley, 1933.

of value in assessing the above soil factors. The salient points in working out the Storie Index of soils are briefly given below.

Four main soil characteristics indicated as A,B,C, and X respectively are studied and evaluated. These are rated with a maximum of 100 points for the most favourable condition and descending to 0 according to the degree of adverse factors operating under field conditions.

Factor A involves a study of the soil profile characteristics under each of the distinct soil groups excluding the texture of the *A*-horizon. Factors in the soil profile which operate adversely under field conditions are involved in the rating under this factor. Thus, in a soil containing a hard pan in its profile, for example, the level at which the hard pan occurs has been used in rating the land for factor A. Rating in this instance has been as follows: hard pan at the depth of 1 foot, 5-20 per cent; at 1-2 feet, 20-30 per cent; at 2-3 feet, 30-40 per cent; at 3-4 feet, 40-50 per cent; and at 4-6 feet, 50-80 per cent.

Factor B relates to the texture of the surface soil. Thus, fine loam, loam, and silt loam are rated at 100, sandy loam at 95, coarse sandy loam and silty clay loam at 90, clay loam at 50 to 70, and gravelly, stony soils at varying ratings from 10 to 20.

Factor C involves the slope of the land. Nearly level (0-2 per cent slope) is rated at 100; gently undulating (0-2 per cent slope), at 95 to 100; gently sloping (3-8 per cent) at 95 to 100; undulating (3-8 per cent) at 85 to 100; moderately sloping or rolling (9-15 per cent) at 80 to 95; strongly sloping or hilly (16-30 per cent) at 70 to 80; steep (30-45 per cent) at 30 to 50, and very steep (45 per cent and above) at 5 to 30.

Under factor X are included all other factors which influence the value of the soil under irrigation, such as drainage, alkali conditions and erosion which are rated from 10 to 100; acidity from 80 to 100; and nutrient supply from 60 to 100. Thus, factor X is proportionately reduced in value from 100 to 0 depending upon the degree of occurrence of each of the above unfavourable factors.

The Index Rating is obtained from the above factor evaluations by a process of multiplications as follows:

$$A \times B \times C \times X \text{ by percentage}$$

The rating value thus obtained may be used to classify land into different groups according to their suitability for irrigation.

Using the above method of ratings, the soils of California in the U.S.A. were grouped under six grades of suitability for irrigated agriculture and the principles adopted in the type of ratings would appear to be suitable for application to conditions obtaining in other parts of the world also.

TABLE XXXVIII

	Index Rating	
Grade I	80-100	Excellent
Grade II	60-79	Good. Suitable for most crops of the region
Grade III	40-59	Fair soils. Suitable for most crops but limited in their use due to presence of undesirable features like hard pan, etc.
Grade IV	20-39	Poor soils. Good for a few crops only
Grade V	10-19	Very poor soils, fit only for pasture
Grade VI	less than 10	Not fit for irrigation

OTHER METHODS OF RATING LANDS FOR IRRIGATED AGRICULTURE

A modification of the above methods has been recently used by a few Indian workers[1] in rating lands for irrigated agriculture in Rajasthan. In this method, the chief soil profile characteristics and the more important physico-chemical properties are given definite marks which taken together total 100, the distribution of the marking being based on the relative importance of each soil characteristic. The characteristics and the maximum marks allotted to each property are as follows:

TABLE XXXIX

Texture	20
Total soluble salts	25
Permeability	20
pH	15
Exchangeable sodium	10
Non-capillary porosity	10
Total	100

[1]K.M.Mehta, C.M. Mathur and H.S. Shankara Narayan, "A Proposed Method for Rating lands for Irrigation and its Application to Chambal Area," *Jour. of Ind. Soc. Soil. Sc.,* Vol. 3. 1958.

The marks are given separately for the soil characteristics for three zones in the soil profile, viz. surface soil, subsoil and substratum. For distinguishing between the three zones, the first soil layer is taken as the surface-soil, then the second layer (and the

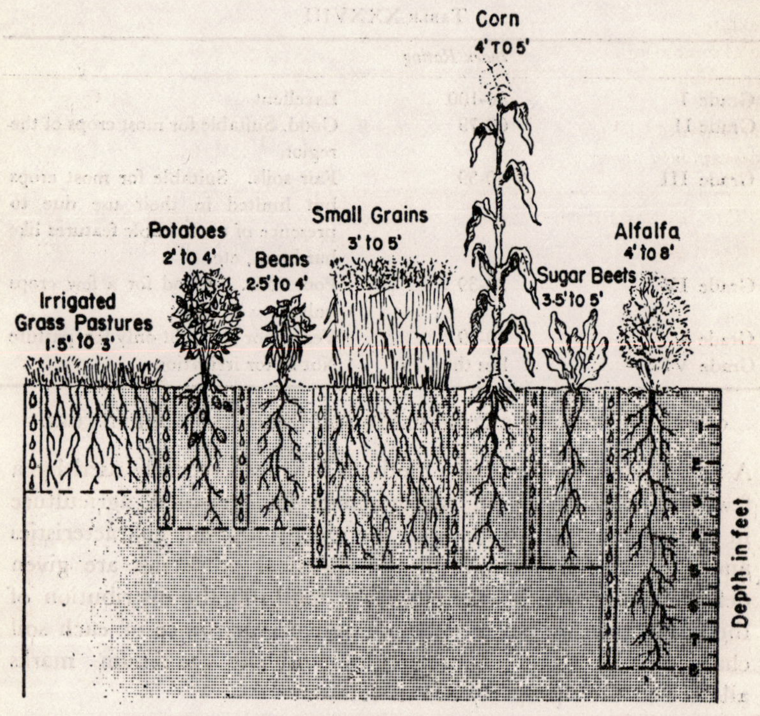

FIG. 99. Different crops can be irrigated to different degrees. Picture shows the depths to which it is recommended to irrigate various crops. The irrigation is controlled to provide water supply adequate to moisten the required soil depth.

third one if more than 3 layers), and the last layer or the last two (when more than 3 layers are present) are taken as the substratum. Surface soil varies from 9 to 12 inches in depth, subsoil up to 4 ft. and substratum from 4 to 6 ft. While assigning marks, the relative importance of each of the soil characteristics in the different zones under irrigation has been taken into consideration.

TABLE XXXX
DISTRIBUTION OF MARKS FOR DIFFERENT ZONES

Soil Property	*Total marks*	*Surface soil*	*Subsoil*	*Substratum valuation*	*Nature of valuation*
Texture	20	8	7	5	Addition
Total soluble salts	25	10	8	7	—do—
Permeability	20	—	13	7	—do—
pH	15	15	15	15	Average
Exchangeable sodium	10	10	10	10	—do—
Non-capillary porosity	10	10	10	10	—do—

The values for the profile characters and their classification are worked out on the basis of the total marks obtained by the various properties.

TABLE XXXXI

Class I	Marks more than 86	Land offering good response to irrigation under usual management practices
Class II	Marks between 76-86	Land creating mild hazards of salinity or permeability
Class III	Marks between 66-76	Land developing impeded drainage, salinity, alkalinity and requiring special management practices
Class IV	Below 66	Land requiring drastic management practices

CLASSIFICATION OF IRRIGATED LANDS AFFECTED BY SALINITY

The presence of soluble salts in soils influence the properties of soils considerably when they are brought under irrigated agriculture. These salts are dissolved and released into the soil solution immediately after irrigation is introduced and are apt to cause adverse effects on the germination of seeds and crop growth. The salts commonly met with in soils are the sulphates, chlorides and carbonates of sodium and calcium, and in rare cases of magnesium. The salt content of these soils is generally expressed as a percentage on the weight of soils or as parts per million. In recent years, these have come to be determined in terms of electrical conductivity of the saturation extract of the soil with distilled water, since electrical conductivity is directly related to salt concentration. Electrical conductivity measurement offers a much simpler method compared to the older method of measuring the actual salt content by extraction of the soil with water and concentrating and weighing the salts. Simple conductivity apparatus based on Wheatstone Bridge principle are now available. The readings from these instruments are expressed either as micro-mhos per centimeter or as milli-mhos per centimeter of soil. The classification of soils based on salt content as followed in the United States of America is given below :[1]

TABLE XXXXII

THE CLASSIFICATION OF SOILS BASED ON SALT CONTENT—LIMITS OF SALINITY CLASSES

Class	*Salt condition*	*Percentage of salt*	*Electrical conductivity of saturation extract in milli-mhos per centimeter*
Class 0	Free	0.0-0.15	0-4
Class 1	slightly affected	0.15-0.35	4-8
Class 2	moderately affected	0.35-0.65	8-15
Class 3	strongly affected	Above 0.65	Above 15

PROBLEMS OF ALKALI SOILS AND THEIR CLASSIFICATION

The presence of sodium in soils in excess of certain limited amounts affect their properties and behaviour, particularly after irrigation. Under these conditions the soils develop alkalinity, associated with a high pH, and deflocculation of the clay takes place and the physical conditions alter in a manner which adversely affects normal cultivation and agricultural operations. The presence of high sodium content in the exchange complex of soils can be detected in the course of pre-irrigation soil surveys and its amount forms the criterion for classification of alkali soils.

The alkali in the soil is measured in terms of exchangeable sodium

[1]*Soil Survey Manual,* U.S. Dept. Agric., Handbook No.18, 1951.

percentage in the soils as determined from the equation :

$$\text{E.S.P.} = \frac{\text{Milliequivalents of Exchange Na/100}}{\text{Cation Exchange capacity of soil}} \times 100$$

The Cation Exchange capacity is defined as the number of milliequivalents of cations such as Na, K, Ca, and Mg that can be held in exchangeable form by 100 gms of soil. A soil having an exchangeable sodium percentage of 15 and more is considered to be an alkali soil. Various laboratory methods are available for the estimation of cation exchange capacity of soils and also the exchangeable sodium percentage but these are generally too elaborate and time consuming to be used for routine analysis in extensive soil surveys involving large numbers of determinations.

SOIL p^H AND ALKALINITY

A relatively simpler estimation of alkali in soils is made by measurements of pH in water suspensions of soils. This method has some limitations and involves some approximations. The measurement of pH, however, is very simple, and can be conveniently measured electrometrically using a pH meter with a glass electrode. It is generally presumed that a soil having a pH of more than 8.5 contains significant quantities of exchangeable sodium. Soils having a pH of over 9.0 have serious alkali problems. In some cases, there may be alkali in the soil even though the soil pH is less than 8.5. This is due to the presence of neutral salts in the soil which depress hydrolysis of sodium salts. Such soils when initially leached or diluted with water will show a gradual rise in pH as the neutral salts are gradually removed from reaction, leaving only the sodium salts. Thus, a soil paste of such a soil may show an initial pH of 8.5 but on dilution to 1 : 5 or 1:10 with water, may show a pH of even 10.0 or 10.5.

OTHER FACTORS FOR STUDY IN PROBLEM AREAS

The depth of the water table is normally an important factor that is to be considered in an irrigated area and this becomes more important if the subsoil waters are saline. The dangers of increase in salinity are great if the saline water table gradually rises on irrigation and establishes capillary contact with the surface. In such a situation, the salts can rise to the soil surface and form deposits there. It is, therefore, useful to make a full study of the depth of the water table and also of the quality of the subsoil water and to prepare a water table map of the area as a part of the pre-irrigation survey.

Further, drainage characteristics, permeability, and the infiltration rates of the soils in the area proposed for irrigation require to be studied since it is very important that water let in after irrigation must move into the soil at predetermined rates. If the infiltration rates and the movement of water are too slow, waterlogging will result. If, on the other hand, these rates are too rapid, they will cause premature localised drying and thus create a need for excess of irrigation waters. Drainage characteristics and permeability can be evaluated by a study of the soil profile, the structure and texture of the subsoil and the presence or absence of hard pans in the profile. Methods for study of infiltration rates depend upon impounding water on the soil surface in metal cylinders of standard dimensions and noting the rate of water-loss through percolation during definite time intervals. Evaporation from the surface of the impounded water during the studies is prevented by floating a film of oil on the water surface. It is considered that a soil with an infiltration rate of less than 0.1 inch per hour is unsuited for irrigation.

REFERENCES

YEGNA, NARAYANA IYER, A.K., *Principles of Crop Husbandry in India*, Bangalore Press, 1957.

SUBBIAH MUDALIAR, V.T., *Principles of Agronomy*, Bangalore Press, 1957.

RANDHAWA, M.S., *Agriculture and Animal Husbandry in India*, I.C.A.R., New Delhi, 1958.

RANDHAWA, M.S., *Agricultural Research in India*, I.C.A.R., New Delhi, 1958.

CHAPTER XV

SALINE AND ALKALI SOILS AND THEIR MANAGEMENT

Under certain conditions of soil development, salts—either neutral or alkaline—can accumulate in the soils. When soils contain high concentrations of salts or excessive amounts of sodium, they develop unfavourable characters which hamper both cultivation operations and growth of crops, and gradually become unproductive. Under extreme conditions they become infertile and barren. These are saline and alkali soils characterised by the presence of high contents of the two types of salts. The distinction between the two types of soils is purely arbitrary, as in many cases, in nature, one may change over to another; or both may occur mixed together; or the same soil may have both the characters. The method of distinguishing these is considered below. Saline and alkali soils may arise either due to natural causes in arid regions, such as in Rajasthan, or in the Coastal plains, as along the West Coast of the peninsula being affected by saline waters of the sea. Or they may develop under man-made conditions due to faulty management and practices in irrigation-agriculture. It is estimated that in India there are about fifteen million acres of such lands distributed in the states of Punjab, Rajasthan, Uttar Pradesh, the Deccan and the Coastal areas. Such lands have been named "Usar," a word derived from the Sanskrit word "Ushtra" meaning barren or sterile. Other local names to describe these soils are *thur*, *kallar*, *karl*, *chowlu* and *chopon*. Besides, the salt-affected soils pose additional problems which require special treatment to make them productive.

ADVERSE EFFECTS OF SALTS ON SOILS AND CROP GROWTH

The bad effects of salts present in the soil on crop growth fall into several categories. They may be either direct, such as happens when chlorides and borates are present in high concentrations and these exert toxic effects on plants, or indirect, such as occurs when pH of the soil increases and depresses the availability of plant nutrients like phosphates, iron, zinc and manganese, and further, affects the biological activities in the soil. The following are some of the important adverse effects on crop growth due to high concentrations of salts in the soil.

(i) *Destruction of soil structure:* The fine crumb structure and soil tilth which are so important for proper crop growth are destroyed under conditions in which there is an accumulation of adsorbed sodium in the soil. This process is known as *Deflocculation.* The soil loses its structure, becomes powdery and on wetting becomes dispersed, sticky and unworkable. Such a soil, when dry, becomes intensely hard and cracks. Thus, plants growing in such a soil will have poor root aeration and are likely to suffer from root injury.

(ii) *High Osmotic Pressure:* Due to the presence of high concentrations of different soluble salts in the saline soils, the osmotic pressure of the soil solution tends to inhibit germination and lowers the power of crop roots to absorb water from the soil. The work of Magistad and Eaton at the Salinity Laboratory, in the U.S.A., has established a linear inverse relationship between the osmotic pressure of the soil solution and the yield of crops. Because of high osmotic pressure in the soil solution, the plants start wilting long before the actual wilting point under normal conditions is reached. Further, the presence of hygroscopic salts such as nitrates and chlorides of calcium and magnesium tends to keep the soil always in a moist and undesirable condition.[1]

RELATIONSHIP OF P^H TO UPTAKE OF PLANT NUTRIENTS

Alkali soils which have high concentrations of exchangeable sodium have a high pH which may range between 8.7 and 10.0. In such soils, phosphate availability becomes highly reduced due to the formation of insoluble forms of soil phosphorus and hence these soils tend to be phosphate deficient. Similarly, under conditions of high pH, soil manganese and boron tend to become unavailable to crop growth. Chlorosis, a yellowing of leaves, is quite a common

[1]Magistad, O.C., and Reitemeier R.F., "Soil Solution Concentrations at the Wilting Point and their Correlation with Plant Growth," *Soil Sci. 53*, 351-360, 1943. Eaton, F.M., "Water Uptake and Root Growth as influenced by inequalities in the Concentration of the Substrate Plant Physical," 16: 545-564, 1941.

occurrence in these soils due to induced iron deficiency. Application of the deficient nutrients to the soil in such circumstances will not be of much use unless the pH is simultaneously brought down by suitable means.

Direct Effects

Some salts when they accumulate in high concentrations, as they do under arid conditions, have toxic effects on plants. Chlorides, for example, have a harmful effect on beans and stone fruits. Borates become toxic to certain plants even when they accumulate in the soil to the extent of about one part per million.

CAUSES OF FORMATION OF SALINE AND ALKALI SOILS

The saline and alkali soils are formed due to certain natural and man-made causes in which undesirable concentrations of salts accumulate in the upper layers of the soils. These salts are mainly chlorides, sulphates, bicarbonates and carbonates (and sometimes nitrates) of sodium, calcium and magnesium (and potassium). In some cases, there is also an accumulation of soluble borates. The origin of these salts can be due to one of two ways. Either these salts are formed *in situ* by a process of gradual weathering of parent materials, which are rich in these elements and have not moved elsewhere, or they have been transported from other places through surface and sub-surface drainage and deposited at these places. Under humid conditions, the salts are either washed down into the lower layers where they get mixed with the ground water and are carried away into streams and ultimately to the ocean or else washed through surface run-off into waterways, streams and rivers. In arid conditions, however, leaching and transportation is not complete and the salts tend to concentrate either on the surface of the soil or at certain depths because moisture is not sufficient to carry them very far. Due to the prevailing conditions of high temperatures and surface evaporation and also high transpiration rates of plants, the salts are carried up to high levels of the soil where they concentrate. While, in some cases, saline soils are directly formed *in situ* by the process of weathering of parent materials, they are often the result of restricted drainage, high water table, impermeability of the soils due to the presence of a hard pan in the surface or sub-surface soil and temporary abundance of humidity interspersed with prolonged dry periods. In India, though salt-affected lands are formed under natural conditions in certain parts of the country, as in Rajasthan and in the Coastal tracts of Gujarat, Maharashtra and Mysore, much of the salt problem has a relation to developments in irrigation agriculture. The areas which come under irrigation usually lie in low elevations and broad plains where drainage is sluggish. Due to injudicious use of water and lack of proper drainage, the water table gradually rises to within five to six feet of the soil surface and establishes capillary contact with it. This results in the rise of salts through the soil solution, and their subsequent deposition on the surface as an efflorescence.

The following is a rough estimate of the extent of saline and alkali soils in the different states in India.

TABLE XXXXIII

ESTIMATED ACREAGE OF SALINE AND ALKALI SOILS IN DIFFERENT STATES

State	*Acreage*
Punjab	3,000,000
Uttar Pradesh	3,100,000
Maharashtra	275,000
Gujarat (Rann of Kutch)	210,000
Rajasthan	1,000,000
North Bihar	10,000
Andhra Pradesh	120,000
Kerala (Kari soils)	38,000
Mysore	1,000,000
Madhya Pradesh	600,000

Source: *Study on Wastelands including Saline, Alkali and Waterlogged Lands and their Reclamation Measures—Planning Commission*, 1963.

EXCHANGEABLE SODIUM PERCENTAGE

The status of an alkali soil is measured in terms of its *exchangeable sodium percentage*. This value is very important in determining the characteristics and behaviour of soils under irrigation and in formulating reclamation measures. The percentage of exchangeable sodium is estimated by the formula:

$$\text{Exchangeable Sodium Percentage (E.S.P.)} = \frac{\text{Milliequivalents exchange sodium/100}}{\text{Cation exchange capacity of soil}} \times 100$$

The cation exchange capacity (c.e.c.) of the soil is the number of milliequivalents of cations held in exchangeable form by 100 gms of soil.

The estimation of exchangeable cations and of sodium is made by standard methods, such as extraction of the soil with neutral normal ammonium acetate solution and estimating the cations in the soil extract by standard laboratory methods.

ELECTRICAL CONDUCTIVITY OF THE SATURATION EXTRACT

A quick and simple method of estimating the total soluble salt content of a soil is to determine the electrical conductivity of its saturation extract. It has been observed that there is a close relationship between the quantity of soluble salts present and the conductivity of the soil extract. In actual practice, it has been found difficult to prepare a filtered saturation extract rapidly enough for large-scale work. Extracts of 1 : 1 or 1 : 2 or even 1 : 5 soil-water ratios are often used for ease of operation, and these determinations are fairly reliable.

For preparing the soil extract, a known quantity of soil varying from 10 gms to 100 gms is taken to make the required ratio, required amount of distilled water is added and the solution is kept stirred for half an hour to attain equilibrium. This suspension is then filtered and the conductivity determinations are made in a Wheatstone-bridge type conductivity apparatus. In some cases, the suspension is allowed to settle for one hour and the clear supernatant liquid is taken for conductivity determinations thus avoiding the filtrations.

The conductivity is expressed in terms of millimhos per centimeter. The millimho is a smaller unit of the mho which is the usual unit for conductivity measurements. The electrical conductivity is expressed by the factor: $EC \times 10^3$ which represents the electrical conductivity in terms of millimhos per centimeter.

For convenience of understanding, because of the more easily described larger figure, conductivity is also expressed in terms of micromhos per centimeter in which case the formula is written as

$$EC \times 10^6$$

The conductivity cell used in these conductivity measurements is of standard type having two platinum electrodes placed parallel to each other, one centimeter apart. The temperature of the solution at the time of measurement and the period of contact of water with the soil require to be controlled to standard conditions, to obtain reliable results.

TYPES OF SALT INFESTED LANDS

Salt infested lands are broadly classified into three categories according to the nature of the salts present. They are:

(i) Saline soils
(ii) Saline-alkali soils
(iii) Alkali soils

The characteristics of each of these classes can be briefly described here.

1. *Saline soils:* These soils contain excessive concentrations of soluble salts which depress the normal germination of seed and the growth of plants. The electrical conductivity of the saturation extract of the soil, which is a measure of the salt content, exceeds 4 millimhos per centimeter. The exchangeable sodium percentage is less than 15. The pH is usually less than 8.5. Adsorbed sodium is present in small concentrations as compared to calcium and magnesium. The principal anions are chlorides, sulphates and small amounts of bicarbonates and nitrates. Soluble carbonates are absent.

Saline soils in Uttar Pradesh are known as *Reh* or *Rehals*, in Punjab as *Thur* and as *Choulu* in Mysore. Extensive white or greyish white fluffy deposits are found on the surface of such soils. There is usually no hard pan or *kankar* in the soil profile. The loss of productivity due to development of salinity in arable land may amount to as much as 10 to 50 per cent. In several cases, the causes of this loss in productivity may not even be recognised by farmers and may be ascribed to other causes.

Soil salinity is also found extensively in the coastal tracts of Gujarat, Maharashtra and Mysore. These are referred to as *Kar* land in Maharashtra and *Jigani* lands in coastal Mysore. These are formed primarily by an inundation of sea water in the form of tides, floods and sprays. These soils are low in productivity.

2. *Saline-alkali soils:* This group has a high concentration of salts and, in addition, has sufficient adsorbed sodium to interfere

FIG. 110. Reclamation of salt-affected lands. The field in the left has been reclaimed by repeated fooding and flushing out of salts. The stand of good crop of paddy is evidence of soil amelioration. The land on the right is still in infertile condition due to salt accumulation.

with the growth of plants. The electrical conductivity of the saturation extract is more than 4 millimhos per centimeter and the exchangeable sodium percentage is more than 15. The pH is usually less than 8.5 depending upon the concentration of the soluble salts. If the concentration of soluble salts goes down as a result of leaching, the relative increase in the sodium percentage in the soil will raise the pH to above 8.5.

The saline soils are called *ret* or *usar* in Uttar Pradesh and *Kallar* in Punjab. Outwardly, the soils may resemble the saline soils, but the subsoil is very hard and more or less impermeable to movement of water. This clay pan may vary in thickness from a few inches to as much as three feet. There may also be a *kankar* pan in the subsoil.

3. *Alkali soils:* These soils contain high exchangeable sodium which interferes with normal plant growth, but have low salt content. The electrical conductivity of the saturation extract is less than 4 millimhos per centimeter. The exchangeable sodium percentage is more than 15. The pH tends to be more than 8.5. When alkali soils occur as small irregular patches in arable land in arid regions, they are called *Slick Spots*. The surface soil is hard and compact and is sticky clay or clay loam with a hard subsoil having a columnar structure. The hard subsoil impedes the free movement of capillary and gravitational water.

When saline and saline-alkali soils are leached to reduce soluble salts with water, without simultaneous applications of lime or gypsum, there is a gradual rise in the pH due to the hydrolysis of exchangeable sodium. Such a hydrolysis leads to the formation of carbonates and to the deflocculation of clay and formation of hard impermeable structures within the soil. The alkali carbonates react with the organic matter present in the soil to form dark coloured organic complexes which give the soil a dark brown colour nd hence are referred to as Black Alkali. Among the soluble cations, sodium is the dominant one, since calcium and magnesium ions are precipitated as insoluble carbonates. The anions are chiefly carbonates with small amounts of chlorides, sulphates and bicarbonates.

Agarwal and Jadhav,[1] based on their own investigations in Uttar Pradesh on the salt-affected soils of the Indo-Gangetic alluvium, have drawn up a salinity and alkali scale to evaluate saline-alkali soils which can be used as a basis for mapping in large-scale soil survey operations. In this method, simple estimations of pH and electrical conductivity of the saturation extract can be used to predict the nature of the crop behaviour.

Alkali soils occur extensively in Uttar Pradesh where it is known as *Usar* and in Punjab under the name *Rakkar* and in the Deccan under the term *Chopan*. In such areas, hard compact soils of clay and clay loam texture lie on heavy compact subsoil with a columnar structure, which is quite impermeable to the movement of water and resistant to the growth of roots.

[1]R.R. Agarwal and J.S.P. Jadhav, *Jour. Ind. Soc. Soil. Sc.,* 1956. Vol. 4, No. 3, pp. 141-5.

Some of the principal characteristics of a saline-alkali *Usar* profile of Uttar Pradesh is given below.

TABLE XXXXIV

SALINITY AND ALKALI STATUS OF A TYPICAL *USAR* SOIL PROFILE OF UTTAR PRADESH

Locality and Local Name	*Hori-zon*	*TSS p.c.*	*Carbo-nate as $Na_2 Co_3$%*	*Bicar-bonate as $Na_2 Co_3$%*	*Chlori-des & Sul-phates of So-dium%*	*pH*	*CEC m.e.%*	*Ex. bases m.e.%*		*Na-satu-ration %*
								Ca	*Na+ K*	
Patchpera	1st	1.61	0.39	0.40	0.74	9.6	10.2	—	10.2	100
Aligarh Dt.	2nd	0.50	0.11	0.16	0.20	9.5	20.76	1.0	19.6	94.4
Usar	3rd	1.23	0.33	0.29	0.58	9.3	15.32	1.0	13.3	86.9
Reh	4th	0.32	0.09	0.14	0.03	9.4	13.44	1.4	11.84	88.1
	5th	0.22	0.03	0.15	0.01	8.2	10.44	2.6	7.04	67.4

MANAGEMENT AND IMPROVEMENT OF SALTED SOILS

There are two important aspects of the salt problem in agricultural soils. One is the reclamation or improvement of natural or man-made salt-affected lands and the other is the prevention of the development of salt concentration in arid and semi-arid regions by the adoption of specially planned management practices.

The two important points to be borne in mind in soil reclamation work are: (i) that the salts are concentrated in certain regions from where they should be removed or rendered incapable of damage, and (ii) that the salts are soluble in water and hence can be transported and removed through water. All the methods that have been drawn up for the improvement of salt-affected lands under irrigation are based on these two fundamental facts. When water is not available in abundance, reclamation programmes are always difficult and slow and can only be imperfect.

Measures for the reclamation of saline-alkali lands are to be based on information obtained on the following points:

1. A proper soil survey and land use mapping of the area and the demarcation of places which are suitable for light irrigation and those that are unfit for irrigation.
2. An investigation into the sources of salinity to find out whether the source of salts is from the minerals present in the parent material or from the irrigation waters or due to transportation from other areas, either in the surface run-off or from the ground water.
3. Determination of the nature and qualities of salts present, pH and electrical conductivity.
4. Study of the soil profile, particularly to note soil texture, structure, occurrence of hard pans in the top soil and subsoil.
5. The natural drainage and ground water levels in the region and the nature and quality of the ground water.
6. Soil management practices in the area.

Based on the above information, reclamation procedures have to be finalised as suitable to the area, taking into consideration the economic factors and the prevailing agricultural practices. The reclamation measures that can be adopted are of several types and may be taken up successively or collectively.

METHODS FOR RECLAMATION OF SALINE SOILS

The first prerequisite for any work of reclamation is the availability of large quantities of water and the provision of adequate drainage. The heavy concentration of salts accumulated in the soil profile can thus be got rid of by systematic flooding and leaching. The affected area is marked off into contour basins with bunds 2 to 4 feet high, which are then flooded with water. The drains are made to a suitable depth, so that the water table is lowered to at least a depth of six feet or more. On flooding, the moisture soaks through the upper layers of the soil and on its passage to lower levels and out into the drains, carries away much of the salts. The number of floodings to be given and the interval between successive ones is dependent upon individual situations. Frequent tests of drainage water may also be necessary to decide the extent of leaching that

FIG. 111. Saline lands are reclaimed by standing water in bunded fields and allowing the waters to drain out. Many areas affected by salinity lack in adequate fall in level to provide for gravity flow of drainage waters. In such cases, provision has to be made for pumping out the salt-charged drainage waters. Picture shows such a pump fitted to a sump to collect saline drainage waters (Imperial Valley Irrigation District, Calif.).

has occurred. According to R.M. Hagan,[1] one foot of water will reduce the salinity of the surface foot of soil by about 80 per cent, and two feet of water will reduce the salinity of the surface soil by 90 per cent and in the second foot by 80 per cent. The leaching procedure is aided if a salt and moisture-loving crop like *dhaincha* or paddy is grown, which will increase soil permeability. It has also been found that drying the soil between periods of flooding is beneficial as cracks develop which will increase the penetration of water. Simple leaching with a provision of drainage for reclaiming saline soils has been found successful in Uttar Pradesh (Dhakauni) and in Punjab in cases where the water table is below 10 feet.

RECLAMATION OF ALKALI SOILS

Reclamation of alkali soils offers greater difficulties than saline soils, because, in addition to leaching soluble salts, the adsorbed sodium is to be replaced by calcium. Here, too, the provision of adequate drainage and leaching measures are necessary. Calcium needed for replacing sodium may be derived from three main sources, viz. the soil itself, or from the leaching, or irrigation water, or application of soil amendments like gypsum, calcium chloride, ground limestone, and other materials containing lime.

In cases where the soil contains enough calcium in the form of *Kankar* nodules or gypsum, advantage may be taken of their presence in the soil in formulating reclamation procedures. The application of acid-forming materials like powdered sulphur, molasses and bulky organic materials, like green manure, paddy straw, paddy husk will release soluble calcium through chemical reactions in the soil, which in their turn will displace the sodium in the exchange complex. When gypsum is present in the soil, the problem becomes simpler, because by a process of opening the soil to expose the gypsum layers and increasing the permeability of the soil, the gypsum may be rendered soluble by flooding.

The procedures described above are rather slow and it may take weeks for the chemical and biological reactions to complete before full leaching operations can be taken up.

[1]Hagan, R. M., 1962. "Some Comments on Irrigation and Drainage as Factors in Saline and Alkali Soils Formation and Reclamation" (presented at the Seminar on Salinity and Alkali Soils Problems, I.A.R.I.).

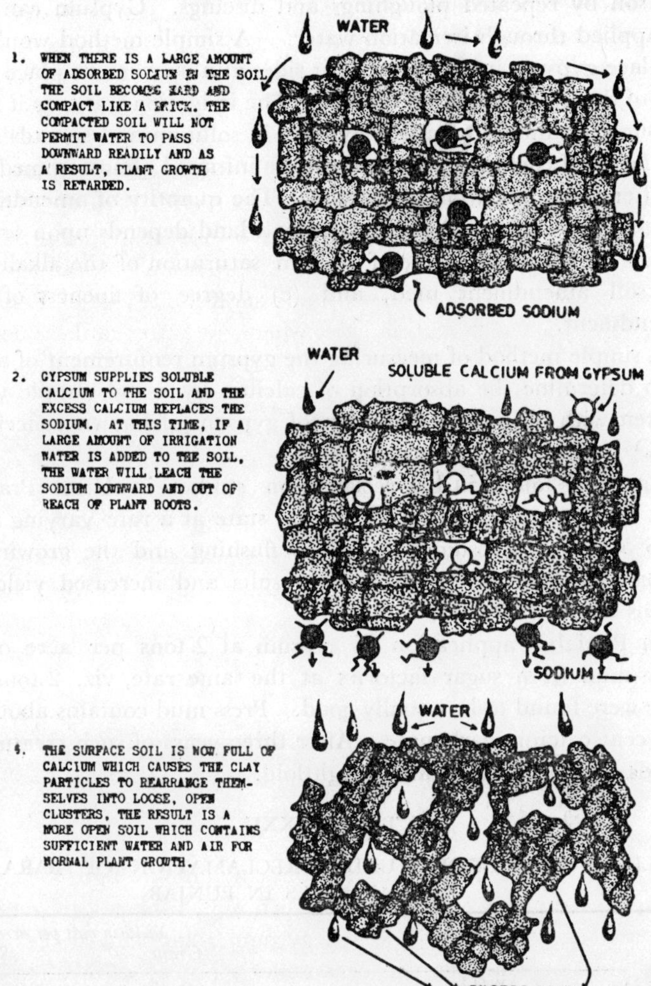

FIG. 112. How gypsum ($CaSO_4$) reclaims alkali soils.

When the soil contains very small quantities of lime, or none at all, the use of soil amendments like gypsum and calcium chloride becomes necessary. Gypsum should be applied in a finely powdered form on the surface of the soil and made to mix thoroughly with the soil by repeated ploughings and discings. Gypsum can also be applied through irrigation water. A simple method would be to place gypsum in a bag with slit sides at a convenient place near one of the main irrigation ditches. The irrigation water as it flows through, gradually takes the gypsum in solution and spreads it on the field. Leaching operations can be initiated almost immediately after the application of gypsum. The quantity of amendments necessary to reclaim a particular alkali land depends upon several factors such as : (a) degree of sodium saturation of the alkali soil, (b) soil amendment used, and (c) degree of fineness of the amendment.

A simple method of measuring the gypsum requirement of a soil is to determine the absorption of calcium by a soil sample when shaken with an aqueous solution of gypsum of known concentration.[1]

Agarwal[2] found in his reclamation studies in Uttar Pradesh that treatment of alkali lands in that state at a rate varying from 3 to 5 tons of gypsum followed by flushing and the growing of *Sesbania aculeata* gave satisfactory results and increased yields of grains nearly threefold.

In Punjab,[3] application of gypsum at 2 tons per acre or of press-mud from sugar factories at the same rate, viz. 2 tons per acre were found to be equally good. Press mud contains about 70 per cent calcium carbonate. After three years of such treatment, yields of rice increased about eightfold.

TABLE XXXXV

YIELD OF RICE GROWN UNDER RECLAMATION OF BARA OR RAKKAR SOILS IN PUNJAB

Year	*Crop*	*Yield in mds per acre*	
		Grain	*Straw*
1933-34	Paddy	2.35	8.2
1934-35	Paddy	7.83	18.72
1935-36	Paddy	19.75	43.12

(Maund=37.324 kgs)

[1]McGeorge, W.T., and Breazeale E.L., "Absorption of Gypsum by Semi-arid Soils," *Ariz. Agri. Exp. Sta. Tech. Bull.* Tucson, 1951.

[2]Agarwal, R.R., 1957, "Alkali Soils," *Ind. Farming*, December.

[3]Nijhawan, S.D., "How to Reclaim Saline and Alkali Soils," *Indian Farming*, 6, No. 7, 1956.

GREEN MANURING, CROP ROTATIONS AND MANAGEMENT PRACTICES

The reclamation of alkali soils is a long, tedious and expensive process. In many cases, the prohibitive cost of soil amendments necessary for reclamation may be beyond the reach of the average cultivator. In such cases, alternative, cheaper but long-term, methods of reclamation are to be adopted. These include levelling, provision of adequate drainage, deep subsoil ploughing to break up hard impermeable pans if any, selection of salt tolerant crops and varieties and suitable crop rotations including green manuring.

The affected areas which are usually at lower levels than the surrounding irrigated or unirrigated lands must be levelled up so that there is a uniform spread of water. Adequate drainage should be provided. In the absence of natural drainage, it may be necessary to dig wells and pump out the water to lower the water table as is currently done in the Punjab. The levelled plots should be bunded, and irrigation or rain-water allowed to stand. The process of infiltration would be quicker if organic matter like paddy straw is applied simultaneously. Flushing with abundant water of good quality is always helpful. Care should be taken that there is no insufficiency of leaching water, especially, if the water itself contains some dissolved salts in it. When the initial leaching has progressed satisfactorily for 3 or 4 months, the land may be planted for a late variety of paddy or a similar crop. The process is repeated during the second year and third year also. Also, crop growth during the first three years may not be satisfactory but the yields will gradually improve from the fourth year onwards. During the fourth year, leguminous green manure crops like *dhaincha* may be grown and ploughed *in situ*. It is expected that by this period the soil is fully reclaimed.

Seed bed

The preparation of seed beds by adopting useful methods is important to ensure successful operations in saline and alkali areas. Seed bed preparation in saline or alkali areas should be done carefully as germinating seeds are especially susceptible to salt injury. Heavy irrigations before and after seeding may be helpful in pushing the salts down, beyond the reach of germinating seeds. In the case of furrow planting as in cotton, the seed should be placed on the side of the furrows instead of at the top of the ridge where the salt concentration tends to be the maximum.[1]

Crop Selection

The growth of salt sensitive crops like potatoes, wheat, *ragi* and beans should as far as possible be avoided on salt-affected land. Salt resistant crops like paddy, sugarcane, barley and vegetables may be grown instead. The change-over has to be made gradually and should be based on local findings and experience. The following classification of some important Indian crops as to their salt sensitivity and salt tolerance has been made by J.S. Kanwar based on his experience in Punjab.

TABLE XXXXVI

Sensitive crops	*Semi-tolerant crops*		*Tolerant crops*	
Field crops	Wheat	Arhar	Barley	Wheat (local variety
(Field beans)	Oats	Flax	Dhaincha	Sugarcane
	Rice	Sunflower	Sugarbeet	Taramira
	Jowar	Castor beans	Tobacco	Salt bush
	Maize		Turnips	Bermuda grass or
			Rape (sarson)	Dub grass (Hariali)
			Cotton	Rhodes grass
Fodders	Turnips	Jowar		
(Guara)	Senji	Maize		
	Sudan grass	Berseem		
	Lucerne	Cowpeas		
	Metha			
Vegetables	Tomato	Onion	Turnips	
(Celery)	Cabbage	Peas	Beetroot	
	Cauliflower	Cucumber	Kale	
	Lettuce	Pumpkins	Asparagus	
	Potato	Karela	Spinach	
	Carrot		Radish	
Fruits	Pomegranate	Orange	Datepalm	
	Olive	Grape fruit	Falsa	
	Fig	Prune		
	Grape	Plum		
	Guava	Almond		
	Mango	Apricot		
	Banana	Peach		
	Pear	Strawberry		
	Apple	Lemon		

[1]Wadleigh, C.H. and Fireman, M., 1949, "Salt Distribution under Furrow and Basin irrigated Cotton and its Effect on Water Removal." *Soil Sci. Amer. Proc.*, 1948, 13.

Based on the experience of a large number of field and laboratory trials conducted in several parts of India over a number of years, the Working Group of the Wasteland Committee of the Planning Commission have drawn up the following procedure for reclaiming different types of salta-ffected lands.[1]

(1) Saline soils with salt efflorescence at the surface.
  (i) By flushing the surface of the soil with water
  (ii) Mechanical scraping of the surface soil.

(2) Saline soils with high concentration of salt at the surface with low water table.
  (i) Ponding the water to leach down the salts.
  (ii) Opening of surface drains and flushing with water to remove salts.

(3) Saline soils with high concentration of salts extending to great depth with high water table.
  (i) Lowering of water table through the use of pumps or wells or tube wells.
  (ii) improving underground drainage through subsoil and tile drains to facilitate removal of excess salts.

(4) Saline soils with hard pan in the subsoil.
  (i) by subsoil or deep ploughing to break the pan mechanically and improve permeability.

(5) Hard alkali soils.
  (i) Using chemical amendments like gypsum, calcium chloride, sulphur, etc.
  (ii) Application of heavy doses of bulky organic manures, like farmyard manures, green manure, molasses, compost, etc.
  (iii) Leaching with water in cases where gypsum occurs in the soil profile.
  (iv) By deep ploughing to remove hard pans in the surface and subsoil and improve permeability.

COASTAL SALINE SOILS

In addition to the inland saline soils associated with arid and semi-

[1]*Study on Wastelands and their Reclamation Measures,* Committee on Natural Resources, Planning Commission, 1963.

arid climatic regions, extensive salt-affected areas occur in India along the coastal lands in the states of West Bengal, Gujarat, Maharashtra, Mysore and Kerala. The area thus affected is estimated to be in the region of eight to ten thousand sq. miles about half of which could profitably be available for reclamation purposes. The source of salinity in these is the sea water which periodically inundates these flat lands during tides and floods, and sometimes is carried in the wind as spray and dust.

The nature of origin and occurrence of the salt lands is different in each area and offers varying problems. The salt lands of Sunderbans in West Bengal have arisen as a result of the inundation by the backwaters of the Bay of Bengal. In Kerala, the saline *Kari* soils along the coastal area, containing a high percentage of organic matter(10-40 per cent), probably formed part of the sea bed previously and are now enclosed in land by alluvial deposits from the rivers.

The coastal saline lands in Gujarat and Maharashtra are called *Khar* and *Khajan*, and in Mysore as *Gajarin*. These are formed due to the periodic inundation of tidal sea water. In Kathiawar, the problem is further complicated by the deposition, over wide areas, of river sediments impregnated with salts. The salts are deposited at varying depths and consist mostly of chlorides (and some sulphates) of sodium and magnesium.

The general plan of reclamation followed in these areas is dependent upon several factors, viz. (1) the nature and extent of salt deposits, (2) the depth at which the salt deposit occurs, and (3) the extent and duration of fresh water available through rivers or rain for the washing out of salts.

The usual procedure is to lay out bunds and embankments on the land to keep out ingress of sea water and to enclose rain-water during the rainy season so that leaching action may take place. Where enough fresh water is available through rivers, dams are constructed across the rivers at higher elevations, and fresh water is ponded on the land periodically to leach out the salts. Drains are provided to carry off the leached out salts.

The reclamation of coastal saline soils requires close cooperation among farmers and the administrative authorities for systematic planning and execution of works and, above all, well scheduled soil management practices so that soils that are once reclaimed at great cost are not allowed to revert to their old state through indifference and neglect.

REFERENCES

GUSTAFSON, A.F., *Using and Managing Soils*, McGraw Hill Book Co., New York, 1948.

U.S. REGIONAL SALINITY LABORATORY, *Diagnosis and Improvement of Saline & Alkali Soils*, 1947.

H. GREENE, "Using Salty Land," *F.A.O. Agricultural Studies*, *No.*3, 1948.

CHAPTER XVI

DRAINAGE OF AGRICULTURAL LANDS

Problems of land drainage involve disposal of surplus water from cultivated lands. Water accumulates in the soil due to one or more causes. It may overflow on surrounding flat country and agricultural lands through storm waters or through river floods during the rainy season or from periodical tides from the sea in coastal regions. Water received during the rains may tend to accumulate in lowlying land due to the locking up of the natural waterways with silt and vegetation. Water may also get collected in the rainy season in old tank beds, in dried-up river beds, in burrow pits along the railway lines and road embankments, and in valleys with restricted outlets. Surplus water in the soil, which collects or stagnates for long periods or continuously may also be caused by a high water table on account of excess of irrigation water or due to seepage losses, from irrigation canals and tank bunds.

Good drainage is necessary for agricultural lands. Movement of water due to normal drainage processes induces aeration and movement of air in the soil profile which is necessary for normal biological growth in the soil. Absence of adequate drainage results in waterlogging, absence of soil aeration, injurious concentrations of salts in the soil and poor productivity. The approximate area of waterlogged lands in different states of India has already been given in Table I and it is seen that 8 to 9 million acres are affected by different types of waterlogging.

Waterlogging either renders the land totally useless for cropping or, if crops are raised, the yields are affected adversely. The problem is, therefore, a large one causing losses in the extent of land available for cropping, and affecting crop production in other ways.

In evolving a suitable drainage pattern for any area, the following points require to be considered. The type of soil, the water table, the amount of annual precipitation received in the region, the quantity of excess water accumulated inside the soil, the type of cropping in vogue in the area, i.e. whether the crops are high-water-demanding or low-water-demanding ones and the sensitivity of the crop to excess soil moisture. In the case of salt-affected lands in arid and semi-arid regions, the drainage should also take into consideration the leaching requirements of the soils.

The amount of water that is to be removed in drainage is expressed in terms of the rate of flow from unit area of land. The quantity of water removed has to allow for the moisture necessary for the normal growth of crops in the area and a reasonably low-level water table for their maintenance and proper growth.

The percolation properties of the soil determine to a large extent the success of any drainage scheme. Soils with poor permeability are difficult to drain and require special measures. The

FIG. 113. In irrigated lands, as in unirrigated lands, provision of drains to lead away drainage waters is necessary for satisfactory cropping.

over-all drainage system is to be designed taking into consideration the general topography of the area and the nature and character of subsoil stratification and the sub-surface layers. Drainage systems for a plain level country will necessarily be different from the methods for drainage in lands with prominent slopes. Where land surface is steep, water movement is fast and advantage of this may be taken while placing drains. Whereas, on level or nearly level land, water movement is sluggish and a larger number of drains may be necessary to remove surplus moisture. Advantage may also be taken of the occurrence of permeable soil layers in the soil profile to place the drains such that the water movement may be quickened.

METHODS OF EFFECTING DRAINAGE

Drainage methods fall into two classes, namely, (a) surface drainage and (b) sub-surface drainage. Surface drainage involves the collection, movement and discharge of the surplus water received on the surface of the land through rainfall and irrigation. Sub-surface drainage deals with the movement and control of water percolating through the soil, either laterally or from the surface down to the water table. Sub-surface drainage controls the depth of the water table.

Deep open drains perform the functions of both surface and sub-surface drainage.

Surface Drains

Surface drains are very necessary in areas receiving high rainfall or where there is large-scale irrigation or where the terrain is flat. In India, surface drainage problems of varying degrees occur in the states of Assam, Punjab, Uttar Pradesh, North Bihar, West Bengal and parts of Orissa and Andhra Pradesh. Surface drains allow free movement of water on the land without causing flooding on cropped land and prevent soil erosion. The surface drains aid in leading the water safely through well-marked channels into natural waterways and streams. The maintenance of drains in general, and surface drains in particular, is important since they are readily susceptible to growth of weeds and grasses and the accumulation of silt, all of which tend to reduce the effectiveness of the drains. It is often observed that natural waterways, which once existed and functioned, are allowed to be silted up and overgrown with weeds and scrub through neglect.

Surface drains on cultivated fields and others, are accomplished in several ways through land formation and providing for even grading and opening of field ditches. Field ditches are either natural or man made. In both cases they require to be cleared of vegetative growth, cleaned periodically and maintained well.

Proper land formation or grading improves surface drainage to a very great extent. The small hollows and lowlying patches of land which may otherwise collect a lot of water, are properly filled up with scraped soil, and the land is properly levelled to allow for free, easy and unobstructed movement of water.

FIG. 114. Field drains are very important for draining away surplus water.

Land formation is necessary especially where the land has very little slope. Land formation may be of different patterns including terracing or formation of bunded plots which are nearly level or of easy grades.

Surface drainage in India is also closely interlinked with the problems created by low-level bridges, road and railway embankments which offer man-made obstructions to the smooth flow of surface flood waters during the rainy season. They often cause pockets of waterlogged land to be formed which are difficult to drain, because the roads or embankments are usually on the ridge lines cutting across the slopes, and in the absence of adequate drains and openings cause obstruction to the free flow of water.

Surface drainage is commonly accomplished in cultivated lands by the formation of field ditches in the land to carry away surplus water. The pattern of field ditches to be adopted in any particular area is dependent upon the topography, type of farming in the area, and the nature of crops to be grown. Field ditches should be so designed as to carry away completely the surface run-off during the wet months without damaging the cultivated land along its course. Ditch systems commonly consist of field drains, lateral ditches, main ditches and drainage outlets. The field drains are usually 9 to 19 inches deep with side slopes, having a ratio of 4 : 1 or 6 : 1. Field drains are located according to topography. Usually, they are found satisfactory if they are placed at a distance of 400 feet in the case of heavier soils and about 1000 feet in sandy soils.

The following important points require to be borne in mind in designing a proper system of field ditches.

1. The ditches should have a gradual slope and be of sufficient depth so that running water may not scour the neighbouring land.
2. The sides of the ditches should be sloping to provide stability. A slope of 1 : 1 in clay soils, and 1 : 2 in loamy soils, and 1 : 3 in sandy soils would be convenient. The ditches should be kept free of vegetation.
3. The ditches should as far as possible run straight or curve gradually. Taking the ditches through sharp curves is to be avoided.
4. The ditches should as far as possible connect all pits and hollows in unformed land and lead them off into streams and natural waterways.

SYSTEM OF SURFACE DRAINAGE

The following five systems of surface drainage are the most common : (i) Parallel system, (ii) Random system, (iii) Bedding system, (iv) Cross Slope system, and (v) Field Ditch system.

Parallel system

The field drains in this method are placed parallel to each other. This is well suited to flat and well formed land. The field drains open into cross ditches which carry off the drainage to natural waterways.

Random system

This consists in connecting hollow places and depressions in the field by a series of lateral ditches. This is done to save cost for levelling of uneven ground. This type of drainage, however, involves waste of considerable land surface due to the opening of numerous lateral ditches, and it also interferes with the normal field operations.

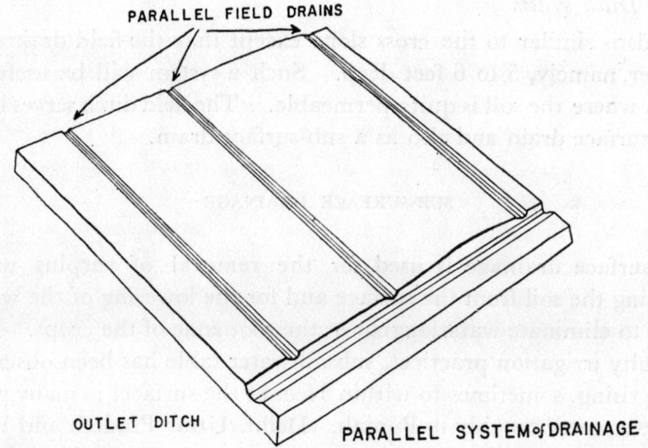

FIG. 115. Diagrammatic representation of the parallel system of surface drains.

Bedding system

This system involves the formation of land into strips or beds in the direction of the slope bordered by dead furrows to carry off surplus water from the beds. Each bed should be smoothed so that there is no hollow or pocket of waterlogged area. The dead furrows open into collection ditches placed across the slope and carry the drainage water into the main drainage outlet. Bedding system is usually used in the row-cropped fields with a small slope of 0.5 per cent or less and with soils of low permeability.

Cross slope system

This is adopted where the soil has a poor vertical drainage. The field drains are formed as in a terrace across a slope at a distance of 100 to 150 feet apart. The field drains open into lateral outlet ditches. The area between the field drains should be levelled and smoothed over.

Field Ditch system

This is similar to the cross slope except that the field drains are deeper, namely, 5 to 6 feet deep. Such a system will be useful in fields where the soil is quite permeable. The field ditch serves both as a surface drain and also as a sub-surface drain.

SUB-SURFACE DRAINAGE

Sub-surface drainage is used for the removal of surplus water entering the soil from the surface and for the lowering of the water table to eliminate waterlogging at the root zone of the crops. Due to faulty irrigation practices, subsoil water table has been observed to be rising, sometimes to within 3 feet of the surface, in many parts of the country notably in Punjab, Delhi, Uttar Pradesh and West Bengal. This has resulted in vast areas getting waterlogged and going out of cultivation. Such areas also suffer in other ways such as in the collapse of houses, bridges and similar structures, rise of the salts to the surface and in affecting the health of the people.

There are several methods of sub-surface drainage, the chief of which are (i) tile drainage, (ii) mole drainage, (iii) perforated pipe or tube drainage, (iv) deep open ditches, and (v) tube-wells.

Tile drainage

In this method, cylindrical hollow tiles are placed in a continuous line at a specified depth and slope grade inside the subsoil. The surplus waters enter the drain system at the tile joint and move out due to gravity.

The tile drains are made of burnt clay or concrete. The former is to be preferred because under certain conditions, such as in the presence of acids or sulphates in the soil, concrete may be attacked. In Punjab, certain locally made tile drains called "Kolabas" have been used with some success. The size of the tiles is varied according to the topography and the quantity of the water that the drain is to carry. The average size would be from four inches to six inches in diameter. The length would be from eighteen to twenty-four inches.

The tile drains are placed inside the soil at predetermined depths by the opening of ditches of required depth. After placing the tiles,

FIG. 116. Provision of field drains is essential for any programme of amelioration of affected saline or alkali lands. Continuously operating machines dig trenches to the required depth and lay the pipes which lead out drainage waters into open ditches.

these are covered over by earth. The system has to be worked out taking into consideration the topography, the soil permeability, the root system of the crops to be grown and the total quantity of water to be transported. There are several systems of tile drainage of which the three important ones are the Parallel System, the Random System and the Interceptor System.

In the Parallel System, the tiles are placed in parallel rows and drain out into a lateral main from one side only. There are several variations of this system depending upon the topography of the land. In the herringbone system which is used when the main or submain lies in a depression, the laterals enter it from both sides, in which case they join the main at an angle.

FIG. 117. Another view of the machine which in a continuous operation after digging the trench at required depth, lays the pipes and covers them over with earth. The pipes are perforated plastic pipes 3″ in diameter and the roll of material is glass wool felt used to roll over the pipes to act as filter and prevent soil from clogging the drain pipe orfices. (Imperial Valley, Irrigation District, Calif.).

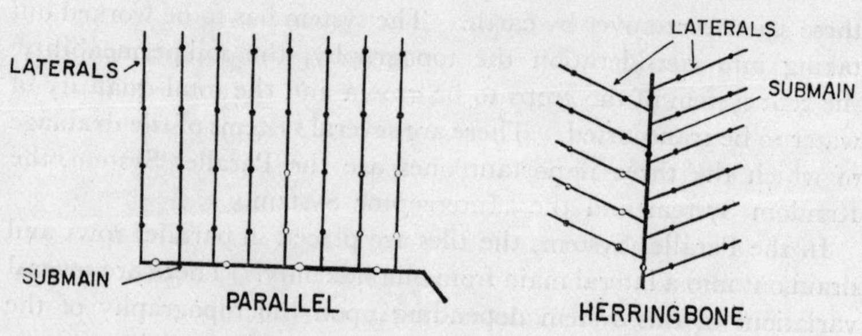

FIG. 118. Systems of providing field drainage by means of tile drains.

The Random System is used in areas where there are scattered wet patches. The usual practice is to lay down the main first, following the surface drainage-wise and then connect the wet areas to mains by placing the laterals.

In the Interceptor System, the drains intercept the seepage water, as it follows an impervious subsoil on its course. The depth for the drainage in this case is determined by taking soil borings to locate the impervious layer in which the drain is placed, with at least a two-feet soil cover to the surface. The seepage water will then flow into the tile drain instead of continuing on its course to cause waterloggoing in the lowlying land further ahead.

The Tile Drain System requires to be provided with air vents to allow for free passage of air and manholes and junction boxes to enable periodical cleaning of the system and keep it free from clogging up.

A good tile drain system requires careful planning and involves considerable expenditure and hence it is not very popular in India. Tile drainage has been used only in restricted areas, notably, in Punjab where it has been found satisfactory in all conditions. However, the economics of the system for large-scale adoption is still under investigation.

Mole drainage

This is similar to tile drainage except that the drainage system has no tiles and hence depends upon the stability of the soil to maintain itself. Mole draining consists in opening a drainage channel by passing a spindle shaped steel mole attached to a vertical blade along the field to be drained. This mechanism is usually attached to a suitable carriage such as a tractor. For the success of the mole drainage, the subsoil through which the drain passes should be uniform undisturbed clay. If there is sand or gravel or pebble pockets, the drains are apt to collapse. Mole drains have been in use in England for more than two centuries. Their average life is short being usually between ten and twelve years. However, their usefulness under Indian conditions is largely unknown.

Perforated metal or plastic pipes of suitable size are sometimes placed in the mole drains to give it stability and in such other cases where the mole drain will not work satisfactorily due to the nature of the underlying strata.

Deep open ditches

These are often employed to lower the water table and act as a surface drainage also. The depth ranges from 5 to 12 feet depending upon topography, soil conditions and the crops grown. However, the disadvantages are that they take away considerable area of the land from cultivation and prove a hindrance for movement of farm equipments and execution of cultural operations.

Tube-wells

In recent years, tube-wells have been used to lower the water table and act as sub-surface drainage in Punjab and Delhi States. The water obtained from tube-wells can with certain limitations be used for irrigation purposes. However, the usefulness of tube-wells for large-scale adoption for subsoil drainage and their economics are still to be worked out.

LINING OF IRRIGATION CANALS

One of the most common causes of waterlogging in soils, particularly irrigated areas, is the seepage of water from the main canals on to the neighbouring lands. It is estimated that nearly 44 per cent of the water let into a canal system is lost in transit before it is received by the crop. Such large-scale seepage of water from irrigation canals has caused vast lands, once fertile, to go out of cultivation in India. This has created serious administrative and social problems and attempts have been made to stop or control seepage from the canals by lining the surface with impervious material. The following materials have been found to be partially successful: (i) *Kankar*, (ii) Cement concrete, (iii) Stone slabs, (iv) Earth compaction, and (v) Bentonite.

The economics and details of operation of such lined canals have still to be largely worked out under actual field conditions. Large-scale experiments have also been made in spraying bitumin on to the surface of the canals to render them impervious to the flowing water and the use of alkathene sheets to serve as impervious layers has been tried. In the areas in the districts of Ganganagar, Jaisalmer and Bikaner where irrigation waters are being led into the desert lands, the losses through seepage are important and the dangers of rise of water table in the adjoining areas, very potent.

Some of the methods indicated above are being tried to offset the dangers of uncontrolled seepage losses, particularly as the rise of water table also poses the problem of the soils becoming saline or alkaline.

REFERENCES

AYRES, C.E. and SCOATES, D., *Land Drainage and Reclamation*, McGraw Hill Book Co., New York, 1939.

STALLINGS, J.H., *Soil Conservation*, Prentice Hall Inc., New York, 1959.

U.S.D.A., *A Manual of Conservation of Soil and Water—Agricultural Hand Book No. 4*, U.S.D.A., S.C.S., Indian Edition, 1964.

CHAPTER XVII

SOIL MANAGEMENT AND FERTILIZER USE IN INDIA

THE maintenance of the schedule of operations in the cultivation of land and cropping in a manner consistent with the requirements of the soil, climate, and cropping conditions, is an important part of management of the soil for optimum production. Besides the various procedures to be adopted and followed in conservation farming, the management of the soil, adoption of manurial treatments and cultivation schedules in keeping with the soil conditions, the season of cropping and the crops themselves are important factors for obtaining the most efficient results.

Soil management in India is closely associated with the following factors: (1) climate of the region, (2) nature of the soil, (3) facilities available for irrigation, and (4) crops grown and the cropping system in vogue.

Taking the above factors into consideration, the following broad regional classification of the different systems of soil management in India can be recognised.

1. Regions having high rainfall—average annual rainfall of over 150 cms, the soil types being of a wide range.
2. Regions having medium rainfall—average annual rainfall ranging from 65 to 150 cms.
   - (i) zones having black soils
   - (ii) zones having red soils
   - (iii) zones having alluvial soils
3. Regions having low rainfall—average annual rainfall of less than 65 cms, with the soil types ranging from the desert type, through red loams to black soils.

REGIONS HAVING HIGH RAINFALL

These areas receive torrential rainfall during the monsoon months ranging between June and November, the annual rainfall being more than 150 cms. They lie in the route of the two principal monsoons in India, viz. the south-west monsoon and the north-east monsoon, in the submontane districts of Himachal Pradesh, Assam, parts of Bihar, W. Bengal and the Western coastal and hilly tracts of the peninsula, in the Maharashtra, Mysore and Kerala States. The soils are usually alluvial, sandy, lateritic or red loams. Soil reaction is generally on the acid side. They are often highly leached and have a low base status. These soils frequently contain high organic matter ranging from 1.5 to 3.0%.

The cropping and soil management pattern in the different soil zones follows their locations and the seasons. Annual crops are confined to the main kharif season and the crop rotations are of a simple nature. Paddy is the main crop which is taken year after year. Sometimes it is rotated with sugarcane, plantains or ginger. Perennial crops like areca and coconuts, and plantation crops, like coffee, tea, pepper and rubber are also grown in regions where the topographic locations are suitable for such cropping.

In the heavy rainfall regions, paddy is the main food crop and this crop may be grown season after season, or year after year, depending upon the rainfall conditions and availability of water for raising this crop under wet conditions. The management of the paddy fields are dependent upon the location of the paddy areas. In the level alluvial tracts of Bengal, Bihar and the eastern parts of Uttar Pradesh, paddy is cultivated in flat bunded fields with provisions for drainage at suitable intervals. In the undulating terrain on the moderately sloping lands in the western regions of the states of Assam, Maharashtra, Mysore and Kerala, the paddy fields are terraced and bunded to hold the rain-water, the natural slope being utilised to provide for drainage. In the more hilly or steeply sloping lands, the terraced fields are more carefully made to provide for better retention of the rain-water, and in these areas, the plots are generally smaller in size and the terracing more intensive to obtain flat stretches of ground where water can be ponded for the growth of paddy.

The preparation of the land for planting paddy has to be done within a very short time immediately after the onset of the first showers in order that the full period of the rainy season may be utilized for successfully going through the growth period of the crop. Hence, short duration varieties are generally preferred in contrast to longer duration varieties which may be grown in the monsoon period in areas of relatively flat terrain and under conditions where supplementary irrigation facilities are available. In the hilly areas where the paddy crop is grown with the monsoon rainfall only, the land is prepared quickly in April or May with the first pre-monsoon showers. With the onset of the monsoon, farmyard manure is spread and the field again ploughed six or eight times. The land is then sown with seed, broadcasting the seeds being a

FIG. 119. Paddy raised under rainfed conditions on bunded fields in terraces. The heavy stand of weeds in fields resulting from inadequate operations. Insufficient retention of the rain-water can also reduce the yield considerably.

common practice. Under conditions where the onset of rains does not permit the sowing of seeds on an extensive scale, the seeds may be drill sown and covered over with soil, a week or so prior to the expected period of regular showers, so that germination begins soon after the ground has become moist. When the plants are about 3 weeks old, one hoeing and one weeding is given. If there is accumulation of rain-water in the fields, another ploughing may be given. In such cases, the seedlings are thinned and, if necessary, replanted in the rows. Such a practice has been known to be more profitable.

Wet cultivation is followed only under assured conditions of water supply. Land is prepared more elaborately than under the "dry" system of cultivation. Two or three ploughings are given and the soil is worked in the wet condition until a fine puddle is formed. Applications of farmyard manure, as well as green manures, either grown in the fields or brought from outside, are made on an intensive scale. Transplanting of 20 to 25 days old seedlings is made in rows in the puddled field. One weeding is made when the plants are 6 to 8 weeks old in the field, and this operation also stirs up the soil to promote better root development.

In the heavy rainfall zone in West Bengal where a crop of jute is raised beginning from the early part of the monsoon, the operations of preparation of the land for paddy, which either follow the jute rotation, or independently, differ somewhat from the procedures indicated above which are largely in vogue in the other areas where paddy is the main crop, with crops like ginger, turmeric and sugarcane taking a place in the rotation.

In the hilly regions and in highly undulating terrain where their cultivation is possible, plantation crops, like coffee, tea, rubber, cardamom and pepper are grown, as also perennial crops, like areca and betelvine. Tapioca and a variety of tuber crops are grown for food under conditions which are favourable for their growth and cultivation. Coffee does best between the elevations of 2500 ft. and 5000 ft. Tea grows over a wider range of elevations. Low grown tea may be cultivated between the elevations of 250 ft. and 2000 ft. at the foot of mountain ranges, with medium elevation teas growing between 2000 ft. and 5000 ft. and the high grown teas cultivated between 5000 and 7000 ft. The quality of the tea grown varies considerably depending upon the elevation at which it is grown, and it is possibly a factor of soil and climatic conditions. Under plantation conditions, soils are better managed than under

FIG. 120. Line planting of paddy enables systematic operations in manuring, weeding and interculture. Line planting is not more labour-consuming than the irregular method of planting and this method is rapidly proving popular in India.

the "dry" field conditions, the higher investment made in such plantation crops necessarily demanding continuous attention to cultural operations, manuring and fertilization besides attention to plant protection and specific needs of the particular crop like pruning, shade provision, etc.

Areca is grown not only on level terrain in elevations under 2000 ft., with irrigation supplementing rainfall, but also in undulating terrain in elevation up to 3000 ft.. The cultivation is dependent purely on rainfall, provision being made to provide for subsoil moisture during the dry part of the year, from water stocked in ponds and small tanks constructed on the upstream side of the hill valleys.

FIG. 121. Areca palms grow at different elevations, but they require plentiful supply of water in all cases.

The soils in the high rainfall regions where crops on plantation scale are raised tend to become acid due to the leaching away of the bases, and liming is regularly resorted to in coffee and areca plantations to keep the pH near neutrality, though slightly on the acid aside.

Tea, however, does well in acid soils and so the soils do not need to be limed in any systematic way. Cultural operations and fertilizer application vary with the crop. For tea, the fertilizer applications are usually made during the spring, while for coffee the doses are split and are applied in three stages—at the pre-blossom, post-blossom and post-monsoon periods. For areca-nut, the manurial applications are made usually towards the end of the south-west monsoon, with lime applications made once in four to six years as necessary. Frequently, banana plantations are grown as pure crop or else mixed with the areca in the high rainfall regions. In these cases, the crop usually gets heavy applications of farmyard manure ranging from 40 to 50 lbs per plant and infrequently, fertilizer applications. In case irrigation is needed, the banana gets irrigation once a week.

REGIONS HAVING MEDIUM RAINFALL

(i) *Zones having Black Soils*

The lands in this region receive from 65 to 100 cms rainfall annually during both the south-west and north-east monsoon periods. The areas with black soil occur in Andhra Pradesh, Madhya Pradesh, Northern Mysore and parts of Gujarat, Maharashtra and Madras. The soils are fertile, rich in bases and lime. They are neutral to moderately alkaline in reaction and are highly moisture retentive. They are, however, low in organic matter, nitrogen and phosphorus. The soil depth varies from shallow to very deep and extends to several feet. Since the soil texture is heavy, being silty clay, clay loams and clayey, they are often difficult to operate when wet. However, a variety of successful crops are grown both under rain-fed as well as irrigated conditions. The principal crops under rain-fed conditions are cotton, wheat, pulses, jowar, oilseeds, maize and tobacco. The crop may either be grown during the *kharif* (south-west monsoon) or during the *rabi* (north-east monsoon) season. Under irrigated conditions, good crops of sugarcane, paddy, cotton, tobacco and wheat are possible.

The cropping season during *kharif* is from June to October, and *rabi* cropping is from October to February. The common practice is to alternate *rabi* wheat with another crop, like maize, cotton, sugarcane, gram, pulses or green manure crops during the *kharif*

TABLE XXXXVII

FERTILIZER RECOMMENDATIONS

Crop	Subregion	Recommended fertilizer dose	Suggested	Fertilizer	Remarks
		N P_2O_5 K_2O	lbs/acre	Grade	
Paddy	Assam	30 : 30 : 15	300	5-10-5	15 lbs N as top dressing
	Andhra Pradesh (Coastal)	30 : 30 : 0	250	6-12-0	—do—
	Bihar	30 : 40 : 20	333	6-12-6	1. Apply at puddling 333 lbs/acre of 6-12-6 grade 2. Top dress with 10 lbs N three weeks after transplanting
	Himachal Pradesh (Transplanted)	33 : 18 : 20	—	—	1. Drill in P and K at sowing time 2. Apply 1/2 N, 3 weeks after transplanting or first timing 3. Rest N, 6 weeks after
	Kerala (Laterite soils)				
	First crop	30 : 30 : 30	300	5-10-5	15 lbs N as top dressing
	Second crop	40 : 30 : 30	300	9-9-9	10 lbs N as top dressing
	Maharashtra	20 : 20 : 0	165	6-12-0	10 lbs as top dressing
	Mysore (Coastal)	30 : 30 : 30	330	9-9-9	10 lbs N top dressing with straight fertilizer
Sugarcane	Assam	120 : 60 : 60	—	—	1. Half of N applied as basal dressing 2. 30 lbs N acre as top dressing at first earthing 3. 30 lbs N acre at 2nd earthing up
Sugarcane	Mysore (Malnad)	150 : 100 : 0	800	5-10-10	120 lbs N with 80 lbs N as basal dose in form of oil cakes
Areca nut	General	50 : 75 : 50			
Cardamom & Pepper	General	50 : 50 : 100			
Plantain	Orissa and West Bengal	30 : 20 : 20	333	9-6.5-9	
	West Bengal				
Jute and Mesta	,,	30 : 20 : 20	333	9-6.5-9	

season. More often, wheat or jowar is sown mixed with other crops like linseed, gram, sarson (Brassica campestris). When a long duration exacting crop like cotton, maize or jowar is grown during

Fig. 122. Crops of linseed and Bengal gram grown in contour strips on black soils in Madhya Pradesh. These are grown in the *rabi* season.

the *kharif* season, the usual practice is to fallow the land during the ensuing *rabi* season.

Fig. 123. Hybrid maize needs plenty of fertilisation and weeding.

TABLE XXXXVIII

FERTILIZER USE

Crop	*Subregion*		*Recommended fertilizer dose*	*Fertilizer mixture suggested*
Wheat	Gujarat	D	20 : 10 : 0	9 : 9 : 0
		I	40 : 20 : 0	9 : 9 : 0
	Maharashtra	D	15 : 0 : 0	—
		I	40 : 20 : 0	12 : 6 : 0
	Madhya Pradesh	D	10 : 10 : 0	9 : 9 : 0
		I	20 : 20 : 0	9 : 9 : 0
	Mysore	D	10 : 10 : 0	—
		I	30 : 30 : 0	—
Jowar	Gujarat		20 : 10 : 0	12 : 6 : 0
	Maharashtra	D	20 : 10 : 0	12 : 6 : 0
	Mysore	I	40 : 20 : 0	12 : 6 : 0
	Kharif		30 : 20 : 0	—
	Rabi		15 : 10 : 0	—
	Irrigated		40 : 20 : 0	—
	Andhra Pradesh		20 : 20 : 20	8 : 8 : 8
Paddy	Gujarat		40 : 20 : 0	9 : 9 : 0
	Maharashtra		40 : 20 : 0	9 : 9 : 0
	Mysore		30 : 30 : 0	—
	Andhra Pradesh		30 : 30 : 30	5 : 10 : 10
Cotton	Gujarat	D	20 : 10 : 0	9 : 9 : 0
		I	40 : 20 : 0	9 : 9 : 0
	Maharashtra	D	20 : 10 : 0	12 : 6 : 0
		I	40 : 20 : 0	12 : 6 : 0
	Madras	I	40 : 15 : 15	8 : 8 : 8
	Mysore	D	20 : 10 : 0	—
			20 : 15 : 15	—
		I	30 : 15 : 15	—
			40 : 20 : 20	—

Source : *Fertilizer Statistics, 1962-63*, Fertilizer Association of India, New Delhi.

D—Rainfed, I—Irrigated

During the *kharif* season, the land is given the barest cultivation. Ploughing for cotton is recommended only once in 3 or 4 years. In other years only the use of a blade harrow is considered adequate for the preparation of land. For pulses, three to four ploughings are given. When preparing land for jowar, one to two ploughings are recommended, followed by the use of a blade harrow.

The black soils can be irrigated for wheat, cotton, jowar, rice and sugarcane. The frequency and quantity are determined by the crop and the season. Wheat is generally given three to four irrigations during the growing season, while cotton and jowar are given occasional irrigation when necessary. Sugarcane is irrigated once in 10 days during the growing season.

A schedule of fertilizer recommendations, made for the various crops and based on the concerned departmental recommendations, is given in Table XXXXVIII.

(ii) *Zones Having Red Soils*

This area comprises parts of Mysore, Madras, Orissa, A.P. and Bihar.

The soils are predominantly clay loams or coarse textured gravelly red loams and have an open texture. The soil depth may vary from a few inches to several feet. The soils are poor in organic matter, nitrogen and phosphoric acid. They have fairly adequate quantities of potash. Soil reaction is generally neutral, but acid or saline soils also occur. Under proper conditions of soil management and water supply, the red soils are highly productive, and satisfactory crops of paddy, sugarcane, vegetables, pulses, groundnut, castor, ragi and other millets, cotton, and a variety of agricultural and horticultural crops can be raised.

The cropping pattern varies according to the *kharif* and *rabi* seasons. During the kharif season, land is prepared for jowar, ragi, groundnuts, maize and other crops. The season lasts from June to November. The land is ploughed immediately after harvesting the crop. If this is not done, it is given thorough ploughing, with 3 or 4 consecutive ploughings followed by the use of a blade harrow during the pre-monsoon showers. Farmyard manure is spread on the land at the rate of 5 to 10 cart-loads (2000 to 4000 kgs) per acre, and then turned into the soil. The seed is either broadcast or drilled in rows, which is more common. Where irrigation

FIG. 124. A high yielding variety of sugarcane. This crop responds well to nitrogenous manuring and detailed schedules of manuring and cultural operations, suited to different soil and irrigation conditions, have been worked out in most states. A stand of crop like the one in the picture yields over 50 tons to the acre.

FIG. 125. Manure heaps distributed systematically over the field. For dry crops addition of farmyard manure or compost ensures supply of nutrients and helps to retain better the moisture received from rainfall.

facilities exist throughout the year, two crops are grown. The rotation followed, in this case, is rice followed by rice, or rice followed by sugarcane, or rice followed by legumes, pulses or millets like ragi. Wherever water supply is limited, as in the case of some minor of irrigation projects or small tank sources as in South India, the land is either left fallow during the summer months, or a catch crop of vegetables or pulse crop, like black gram or horse gram may be taken utilising the residual moisture in the soil.

Good crops are obtained in the red soils by a system of balanced and judicious fertilizer application. Table XXXXIX lists some of the recommended doses in different parts of India for the principal crops.

(iii) *Regions having alluvial soils*

These regions lie in Punjab, Uttar Pradesh, West Bengal, Madhya Pradesh, Bihar and the coastal districts of Andhra Pradesh and Madras. The alluvial soils have been deposited extensively in India by the major rivers like the Ganges, Indus, Godavari, Kistna, Cauvery and their tributaries. The soil depth is variable and often extends to several feet. Since these are transported soils, they bear no intimate relation to the rocks occurring below. Profile development may not be distinguished by distinctive horizons but they may show alternating layers of silt, sand and clay. They are rich in potash and may contain lime concretions either distributed uniformly in depth or as *kankar* layers. They are, however, poor in organic matter and phosphoric acid. They are considered to be highly productive and bear successfully a variety of agricultural and horticultural crops. However, under defective drainage conditions, such as occur in Punjab and parts of U.P., they tend to become saline and unproductive.

In most areas the soils are cropped either singly or as mixed crops. During winter, the crops are wheat, linseed and gram and during *kharif* rice, jowar, cotton or maize, or sometimes the fields are left fallow. In the rabi season, the surface of the soil is opened up with a *desi* plough, five or six times prior to levelling and sowing. The deeper layers are left undisturbed in order to conserve moisture. If irrigation facilities are available, one ploughing with a furrow-turning plough is given. Sowing is done either broadcast or in rows with a drill. Generally, a single weeding and interculturing may be given a little time after germination. Where perennial irrigation facilities are available, rice, sugarcane, peas and other vegetables are grown. Rice cultivation follows the usual wet cultivation pattern wherein the land is thoroughly puddled before transplanting the seedlings. Fertilizer use gives good responses in this area. Table XXXXIX gives the general fertilizer recommendations in the different states for the principal crops.

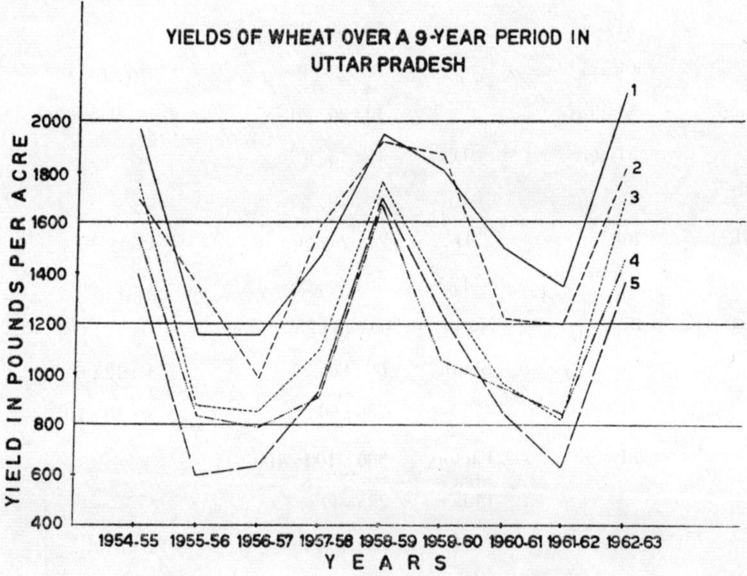

FIG. 126. Wheat yields were highest with the continued use of only chemical fertilisers than with the use of farmyard manure plus chemical fertilisers.

REGIONS HAVING LOW RAINFALL

These regions receive less than 65 cms rainfall yearly and the rains are also ill-distributed. Such areas occur in Rajasthan, parts of Maharashtra, Gujarat, Mysore, Madras and Andhra Pradesh. The soils in these areas are of various types. Soils in Rajasthan are wind-blown grey and brown desert sands with low organic matter but with fair amounts of phosphates and lime. The soil reaction is on the

TABLE XXXXIX
FERTILIZER USE

Crop	Subregion		Recommended Fertilizer dose	Fertilizer mixture recommended
Paddy	Bihar	D	25 : 30 : 30	6 : 12 : 6
		I	50 : 40 : 20	6 : 12 : 6
	Madras		30 : 30 : 0	6 : 12 : 6
	Mysore		30 : 30 : 20	———
	Orissa		30 : 20 : 0	10 : 10 : 0
Ragi	Madras		40 : 20 : 20	8 : 8 : 8
	Mysore	D	15 : 15 : 15	———
		I	30 : 30 : 30	———
Wheat	Bihar	D	25 : 25 : 20	6 : 12 : 6
		I	50 : 40 : 20	6 : 12 : 6
Sugarcane	Bihar	North	60 : 75 : 25	6 : 12 : 6
		South	80 : 60 : 25	6 : 12 : 6
	Madras		250 : 60 : 80	5 : 10 : 10
	Mysore	Factory area	300 : 100 : 100	———
		Other areas	225 : 175 : 75	———
	Orissa	D	80 : 40 : 40	8 : 8 : 8
		I	120 : 60 : 60	8 : 8 : 8
Potato	Bihar		60 : 40 : 40	8 : 8 : 8
	Madras		120 : 200 : 100	5 : 10 : 5
	Mysore		50 : 25 : 25	———
	Orissa		60 : 40 : 40	8 : 8 : 8

N.B. D—Rain-fed
I—Irrigated

alkaline side, ranging from pH 7.5 to 9.2. The salt content is generally high.

FIG. 127. Ragi (Eleucine coracana) is an important rain-fed crop of the Mysore plateau. The crop which can be raised under irrigated conditions also, yields over 2000 kg per hectare.

The soils in Maharashtra, Andhra Pradesh, Gujarat, Mysore and Madras are predominantly black, interspersed with sandy red loams. Here, too, organic matter content is poor and the soils have low nitrogen and phosphate status. Soil reaction tends to be on the alkaline side.

Management of soils in the low rainfall regions is mainly concerned with proper soil management and water conservation measures, as shortage of moisture in the soil forms the principal limiting factor. In parts of Maharashtra, Punjab and Mysore, improved methods of dry farming are recommended. The following are the salient features of the dry farming technique evolved for adoption in the areas after extensive field investigations in the Dry-farming Research Stations at Bijapur, Sholapur, Rohtak and Hagari.

1. Ploughing once in six years in black soils of Bombay and Mysore are enough. Three harrowings and one inter-culturing every year have proved quite effective for jowar. This will keep

loss of soil moisture to the minimum.

2. The fields are to be contour-bunded to prevent surface run-off of rain-water and check erosion of fertile top soil.

3. Lower seed rate and wider spacing are to be given to facilitate better root and better absorption of soil moisture.

4. Balanced manuring and fertilization to keep soil productivity at optimum level.

5. Strip-cropping with alternate strips of erosion resisting crops (pulses and other legumes) and erosion permitting crops (cereals).

6. Adoption of rotation of cereals with legumes to conserve soil moisture and fertility levels.

7. Fallowing the land after taking an exacting crop like jowar or cotto.n

FIG. 128. Two varieties of ragi (eleucine coracana) show characteristic differences. The one on the right has closed compact heads while the other one is of the open type.

Use of fertilizers in dry-farming areas is minimal on account of the uncertainty of rains and hence of the crop itself. However, if there is an assurance of good rains, limited applications of fertilizers are recommended. Table L details the fertilizer dosages recommended for crops in the low rainfall regions.

TABLE L
FERTILIZER USE

Crop	*Subregion*	*Fertilizer dose*	*Fertilizer mixture recommended*
Wheat	Punjab		
	Hissar	4 : 20 : 20	—
	Kangra	40 : 40 : 40	—
	Bhatinda	40 : 20 : 0	—
	Other districts	40 : 0 : 0	—
	Uttar Pradesh		
	D	10 : 10 : 0	9 : 9 : 0
		15 : 15 : 15	9 : 9 : 0
	I	20 : 20 : 0	9 : 9 : 0
		30 : 30 : 0	9 : 9 : 0
	West Bengal	30 : 20 : 0	—
Rice	Andhra Pradesh	30 : 15 : 0	10 : 10 : 0
	Punjab		—
	Gurdaspur		
	Ferozepur		—
	Patiala	40 : 20 : 20	
	Kangra	30 : 30 : 30	—
	Other districts	30 : 0 : 0	—
	Uttar Pradesh	45 : 23 : 0	9 : 9 : 0
	Madhya Pradesh	10 : 10 : 0	
	D	20 : 20 : 0	6 : 12 : 0
	I	40 : 40 : 0	6 : 12 : 0
	West Bengal	25 : 20 : 0	10 : 9 : 0
Sugarcane	Punjab	60 : 30 : 30	—
		100 : 0 : 0	—
	Uttar Pradesh	135 : 67 : 0	6 : 12 : 0
	Madhya Pradesh	150 : 100 : 0	6 : 12 : 0
	West Bengal	120 : 0 : 0	
		100 : 70 : 0	—
	Andhra Pradesh	150 : 50 : 0	10 : 10 : 0
	Punjab	100 : 50 : 50	—
Maize	Uttar Pradesh	135 : 67 : 0	—

TABLE LI

FERTILIZER USE

Crop	*Subregion*	*Fertilizer dose*	*Fertilizer mixture*
Jowar	Rajasthan	40 : 20 : 0	9 : 9 : 0
	Maharashtra	20 : 10 : 0	12 : 6 : 0
	Mysore	15 : 10 : 0	———
	Andhra Pradesh	20 : 20 : 20	8 : 8 : 0
Bajra and	Rajasthan	15 : 15 : 0	———
Small millets	Maharashtra	10 : 0 : 0	———
	Mysore	15 : 10 : 0	———
	Andhra Pradesh	20 : 20 : 20	8 : 8 : 8

NEW STRATEGY IN CROP PRODUCTION—MANURING OF HYBRID VARIETIES

The development of hybrid varieties, which respond to fertilization with high levels of nutrients compared to the traditional varieties, has opened up fresh possibilities for use of larger doses of fertilizers. Their qualities of taking up more nutrients with increased crop production without putting forth much vegetative growth combined with their relatively shorter growth which has earned for them the appellation of "dwarf" varieties, enables such varieties to grow to full maturity without lodging. Thus the hybrid varieties of maize, bajra and wheat and also the exotic varieties of paddy, have been found to respond to high dosages of N, P, and K and to produce high yields of crops. In view of those qualities these varieties are being popularised extensively in the country in the new strategy for increased crop production.

Below is given a brief abstract of the manurial requirements of the improved varieties of these different crops and also their responses.

TABLE LII

Crop	*Optimum recommended dose in kg per hectare*	*Response (increase over no fertilizer) in quintals/hectare*
Wheat		
Tall—improved variety —C 306	80 N 40 P_2O_5 40 K_2O	20
Dwarf Sonora 64 and Lerma Rojo	160 N 40 P_2O_5 40 K_2O	45
Maize		
Hybrid maize	120 N 50 P_2O_5 50 K_2O (plus zinc sulphate at 5 kg per ha)	25
Bajra		
Pusa Moti (Local) (Barani—rain-fed)	30 N 20 P_2O_5	5
Hybrid bajra (Irrigation supplemented)	60 N 40 P_2O_5	20
Paddy		
N. P. 130 Tall	60 N 40 P_2O_5 40 K_2O	
Dwarf Taichung Native 1	150 N 40 P_2O_5 40 K_2O	50

FIG. 129. Field trials are important means of ascertaining the behaviour under actual cropping condition of different varieties under similar treatments, and the effects of different manurial and other treatments on crops.

It is easy to recognise that the additional input in terms of the cost of fertilizers applied is more than compensated by the larger outturn of crop, and this even after taking into account the increased measures of plant protection is needed to safeguard the larger crops.

HIGH YIELD PERFORMANCE OF DWARF WHEATS

The potentialities of high yield performance shown by dwarf wheat varieties induced comprehensive agronomic research in the years following 1963-64. Comprehensive and coordinated study of the agronomic practices and manurial requirements of these varieties were taken up on country-wide basis and detailed information on their cultural requirements, optimum date of sowing, seed rate and spacing, as also the methods and depth of sowing have been worked out. Detailed information on these are available from a bulletin of the Indian Agricultural Research Institute entitled "Five years of Research on Dwarf wheats" recently brought out.

Based on these trial responses to graded doses of nitrogen at 40 kg intervals, starting from 0-40 kg N per ha and going upto 200 kg N per ha have been worked out on 11 varieties of dwarf wheat and the responses to N and also the net profits on the increased fertilization are given below.

TABLE LIII

OPTIMUM DOSES OF N (KG/HA) AND PROFITS FROM FERTILIZATION FOR DIFFERENT WHEAT VARIETIES*

Variety	Optimum dose of N (kg/ha)	Yield at optimum (kg/ha) (kg/ha)	Response over no fertilizer (kg/ha)	Net Profit (Rs./ha)	Net profit per rupee spent on fertilizers (in rupees)
Sharbati Sonora	124.0	4415	2225	1532	6.2
S-277	95.2	3949	1571	1066	5.6
S-308	105.2	3694	1475	970	4.6
PV-18	95.2	3821	1441	962	5.1
Chhoti Lerma	105.2	3817	1631	1094	5.2
Sonora-63	84.4	3291	840	503	3.0
Sonora-64	118.0	3888	1748	1162	4.9
Lerma Rojo	92.0	3596	1221	793	4.3
J-277	92.8	3652	1172	752	4.0
Safed Lerma	89.6	3422	1223	799	4.5
C-306	69.6	3154	785	489	3.5

*It has been assumed here that the cost of kg. of Nitrogen is Rs. 2.00 and that of a kg. of wheat Rs. 0.80, corresponding to a price ratio of 2.5.

It will appear from the above that five varieties have shown a positive response to nitrogen beyond a dose of 120 kg N/ha. Sharbati Sonora has shown the best response to nitrogen at all levels up to 160 kg/ha followed by Sonora 64.

FERTILIZER RECOMMENDATIONS—HIGH YIELDING VARIETIES

The high yielding varieties programme sponsored by the Agricultural Departments of the various states in the country has caught the imagination of the farmers and in the past few years there is a rapidly increasing trend of farmers towards adopting intensive cultivation practices under this programme. The farmers require guidance in regard to the fertilizer dosages to be adopted for the high yielding varieties programme and also the Intensive Agricultural District programmes and various States have drawn up detailed schedules to fulfil this need. A summary is made below of the fertilizer recommendations in the high yielding varieties programme in various States and this information would be found useful.

TABLE LIV

FERTILIZER RECOMMENDATIONS
HIGH YIELDING VARIETIES PROGRAMME

State	*Crop & variety*	*Recommended Rate (lbs/acre)*	*Instalments*
1	2	3	4
		N P K	
Gujarat	Paddy: Formosa 3 T.N.1	88 : 66 : 33	First dose 33 : 33 : 33 before planting or sowing
			Second dose 33 : 33 : 0 to be applied 20 days after planting
			Third dose 22 : — : — after 45 days of planting
	Wheat: NP-824 Mexican.	88 : 44 : 33	first dose: 33 : 33 : 3 before sowing of crop
			second dose 22 : 11 : — after 15-20 days of sowing
			3rd dose 33 : — : — after 45 days of sowing
	Jowar Hybrid Bajra Hybrid and Cotton	66 : 33 : 33	1st dose 33 : 33 : 33 before sowing
			2nd dose 33 : — : — after 15 to 20 days of sowing
	Hybrid Maize: Ganga-3 Ranjit	110 : 88 : 44	1st dose 44 : 44 : 44 before sowing
			2nd dose 44 : 44 : 0 after 15 days of sowing
			3rd dose 22 : 0 : 0
Haryana	Wheat (irrigated)	44 : 20 : 20	Drill 1/2 N, P & K at sowing time and remaining 1/2 N by broadcast with first irrigation. (For C.273 variety application of K_2O is essential.)
	Mexican	120 : 60 : 60	—do—

TABLE LIV (*continued*)

1	2	3	4
	Maize Hybrid (all districts)	100 : 50 : 50	Drill P, K and 1/2 N at sowing and apply remaining N after 1 month as side dressing
	Paddy (Karnal Ambala).	40 : 20 : 20	Apply P & K at puddling, 1/2 N three weeks after transplanting and remaining N, 6 weeks after transplanting
Haryana	Paddy, T.N. 1 IR-8 (all districts)	120 : 60 : 60	Drill in 1/2 N and all P and K at puddling. Apply rest N each at interval of 20 days after transplanting in two split doses
Madras	Paddy : ADT-27 Co-25 Co-29	62 : 46 : 31	Apply 1/2 N, P and K as basal dressing (g.m. 5000 lbs/acre also) Remaining 1/2 N as top dressing
	*others	120 : 60 : 60	*Basal dressing of 15 : 31 : 15 and remaining 25 lbs N as top dressing
Maharashtra	Paddy TN-1	88 : 33 : 22	1. first dose 66 : 33 : 22 to be applied at puddling 2. Second dose of 22 lbs N to be given 4 weeks after planting
	Hybrid Jowar (assured rainfall)	44 : 22 : 11	1. First dose 22 : 22 : 11 to be applied at sowing 2. Remaining 22 lbs N to be given 25-40 days of sowing
	Hybrid Jowar (under irrigation)	66 : 44 : 33	1. First dose 33 : 44 : 33 applied at sowing time 2. Remaining 33 lbs N applied 25-30 days afterwards
	Hybrid Maize	88 : 55 : 44	1. First dose of 44 : 55 : 44 applied at sowing time

TABLE LIV (*continued*)

1	2	3	4
			2. Second dose of 44 lbs N applied 4 weeks afterwards
	Mexican Wheat	110 : 55 : 44	1. First dose of 66 : 55 : 44 applied before sowing 2. Second dose of 44 lbs N given 3 weeks after sowing
Mysore	Paddy Taichung (all districts)	80 : 40 : 40	444 lbs of 9-9-9 and 40 lbs N as straight fertilizer
Punjab	Maize-Hybrid	100 : 50 : 50	Drill P, K and 1/2 N at sowing and apply remaining N after one month as side dressing
	Paddy T.N.-1 IR-8	120 : 60 : 60	Drill P, K and 1/2 at puddling. Apply rest of N each at intervals of 20 days after transplanting in 2 split doses
Uttar Pradesh	Paddy T.N. 1 (all districts)	80 : 40 : 40	Organic manure 30-35 mds/acre
	Local high yielding N. S. T. 98/200/ch. 4	60 : 30 : 30	—do—
	Maize (1) Hybrid (all over state)	80 : 40 : 40	—do—
	(2) Ganga 101/Type 41/Jaunpur	60 : 30 : 30	—do—
	Jowar and Bajra Hybrid	60 : 30 : 30	—do—

REFERENCES

ARAKERE, CHALAM, SATYANARAYANA AND DONAHUE, *Soil Management in India*, Asia Publishing House, Bombay, 1959.

FERTILIZER ASSOCIATION OF INDIA, *Fertilizer Statistics*, New Delhi, 1964-65.

SUBBIAH MUDALIAR, V.T., *Principles of Agronomy*, Bangalore Press, 1957.

YEGNANARAYANA IYER, A. K., *Principles of Crop Husbandry in India*, Bangalore Press, 1957.

APPENDIX

LOCATION OF THE REGIONAL OFFICES OF ALL INDIA SOIL AND LAND USE SURVEY

1. Chief Soil Survey Officer,
   All India Soil and Land Use Survey,
   Indian Agricultural Research Institute,
   New Delhi-12.

2. Soil Correlator,
   Delhi Regional Centre
   All India Soil and Land Use Survey,
   Indian Agricultural Research Institute,
   New Delhi-12.

3. Soil Correlator,
   Nagpur Regional Centre
   All India Soil and Land Use Survey,
   Chindwara Road,
   Nagpur-1.

4. Soil Correlator,
   Calcutta Regional Centre
   No. 233, Netaji Subhas Chandra Bose Road
   Tollygunge,
   Calcutta-47.

5. Soil Correlator,
   Bangalore Regional Centre
   All India Soil and Land Use Survey
   Hebbal,
   Bangalore-24.

TABLE 1A

LOCATION OF SOIL TESTING LABORATORIES IN INDIA

State	*Location*	*Address*
Andhra Pradesh	Bapatla	College of Agriculture, Bapatla.
	Hyderabad	Main Farm, Rajendra Nagar, Hyderabad Dn.
	Rajahmundry	Central Tobacco Research Institute, Rajahmundry.

TABLE 1A (*Continued*)

State	*Location*	*Address*
Assam	Jorhat	Department of Agriculture, Jorhat, Gauhati, Silchar.
	Hazaribagh	Soil Conservation Department, Damodar Valley Corporation, Hazaribagh.
Gujarat	Junagadh	Department of Agriculture, Junagadh.
Kerala	Trivandrum	Agricultural College, Vellayani, P.O. Nemon, Trivandrum.
Madhya Pradesh	Gwalior	College of Agriculture, Gwalior.
	Jabalpur	Agricultural Research Institute Adhartal, Jabalpur.
Madras	Coimbatore	College of Agriculture, Lawley Road, P.O. Coimbatore.
Maharashtra	Poona	College of agriculture Poona.
Mysore	Bangalore	Dept. of Agriculture, Bangalore.
	Mysore	First Floor, 2638 II Main Vanivilas Mohalla, Mysore-2.
Maharashtra	Sambalpur	Agricultural Research Station, Sambalpur.
Punjab	Karnal	C-149 Model Town, Karnal.
	Ludhiana	College of Agriculture, Ludhiana.
Rajasthan	Jodhpur	Department of Agriculture, 24 Paots, Jodhpur.
Uttar Pradesh	Kanpur	College of Agriculture, Kanpur.
West Bengal	Calcutta	Department of Agriculture, 230, Netaji Subhas Chandra Road, Calcutta-40.
Delhi	Delhi	Division of Chemistry, Indian Agricultural Research Institute, New Delhi-12.
Himachal Pradesh	Simla	Department of Agriculture, Hawthorn Villa, Simla-4.
Tripura	Agartala	Department of Agriculture, Agartala, Tripura.

Table 2A

ALL INDIA SOIL TEST SUMMARIES AND NUTRIENT INDEX VALUES (STATEWISE) UP TO 31-12-1962.

State/Territory	*Available Nitrogen*					*Available Phosphorus*					*Available Potash*				
	No. of soils	*Percentage L*	*M*	*H*	*Nutrient index*	*No. of soils*	*Percentage L*	*M*	*H*	*Nutrient index*	*No of soils*	*Percentage* L	*M*	*H*	*Nutrient index*
Andhra Pradesh	59,583	41	57	22	1.81	60,000	47	31	22	1.75	196,969	8	33	59	2.18
Bihar	13,969	64	23	13	1.49	16,774	64	22	14	1.50	7,613	23	49	28	2.05
Delhi	24,492	76	18	6	1.30	24,469	43	38	19	1.76	22,500	8	48	44	2.36
Gujarat	779	48	30	22	1.74	972	52	34	14	1.62	—	—	—	—	—
Himachal Pradesh	18,223	31	64	5	1.74	18,223	16	53	31	2.15	18,223	18	60	22	2.04
Kerala	32,227	19	73	8	1.89	32,228	70	19	11	1.41	32,228	76	20	4	1.28
Madhya Pradesh	19,678	70	19	11	1.41	29,414	37	28	35	1.98	20,859	9	30	61	2.52
Madras	20,938	86	12	2	1.16	20,599	61	25	14	1.53	—	—	—	—	—
Maharashtra	32,140	57	29	14	1.57	34,094	66	25	9	1.43	10,129	2	26	72	2.70
Mysore	14,879	50	23	27	1.77	14,908	80	17	3	1.23	14,856	28	36	36	2.08
Orissa	15,509	53	31	16	1.63	15,663	54	29	17	1.63	—	—	—	—	—
Punjab	35,299	50	27	23	1.73	28,844	45	39	16	1.71	902	52	35	13	1.61
Rajasthan	22,245	79	17	4	1.25	21,546	42	34	24	1.82	13,621	15	58	27	2.12
Tripura	759	36	53	11	1.75	2,044	54	32	14	1.60	—	—	—	—	—
Uttar Pradesh	3,299	89	10	1	1.12	3,056	37	38	25	1.88	3,620	2	88	10	2.08
West Bengal	14,814	49	38	13	1.64	23,555	50	40	10	1.60	9,039	17	47	36	2.19
All India	328,833	54	33	13	1.59	346,389	51	31	18	1.67	373,286	25	40	35	2.10

L=Low M=Medium H=High

TABLE 3A

LAND UTILIZATION IN INDIA (STATISTICS FOR 1958-59)

	Total Area and Classification of Area	*(Thousand acres)*
Total geographical area	According to professional survey	806,270
	According to village papers	726,111
Not available for cultivation	Land put to non-agricultural uses	33,501
	Barren and unculturable land	81,213
	Total	114,714
Other Uncultivated land excluding fallow land	Permanent pastures and other grazing lands	32,386
	Land under miscellaneous tree crops and groves not included under net area sown	14,105
	Culturable waste	50,907
	Total	97,398
Fallow lands	Fallow lands other than current fallows	30,347
	Current fallows	29,423
	Total	59,770
	Net area sown	324,123
	Total cropped area	372,762
	Area sown more than once	48,639

Source: Agricultural Situation in India, No. 1962, Vol. 17, No.8.

TABLE 4A

NET CULTIVATED AND IRRIGATED AREA IN THE STATES, 1958-59

State	*Total reporting area (thousand acres)*	*Net cultivated area*	*Percentage of cultivated area to total area*	*Net irrigated area (thousand acres)*	*Percentage of irrigated area to cultivated area*
Andhra Pradesh	67,748	27,302	40.0	7,046	25.8
Assam (including NEFA)	35,764	5,118	14.3	1,533	30.0
Bihar	42,823	19,690	46.0	5,090	25.9
Maharashtra and Gujarat	121,317	67,438	55.6	3,807	5.6
Jammu & Kashmir	5,927	1,597	26.9	736	46.1
Kerala	9,535	4,587	48.1	879	19.2
Madhya Pradesh	108,376	38,786	35.8	2,072	5.3
Madras	32,021	14,326	44.7	5,586	39.0
Mysore	46,275	25,142	54.3	1,922	7.6
Orissa	38,401	13,854	36.1	2,414	17.4
Punjab	30,284	18,488	61.0	7,361	39.8
Rajasthan	84,300	31,104	36.9	3,571	11.5
Uttar Pradesh	74,141	42,122	56.8	12,137	28.8
West Bengal	21,874	12,929	59.1	3,339	25.8
Delhi	366	217	59.3	79	36.4
Himachal Pradesh	3,882	667	17.2	96	14.4
Manipur	346	232	67.1	149	64.2
Tripura	2,634	500	19.0	8	1.6
Andaman & Nicobar Islands	90	17	18.9	—	—
Laccadive, Minicoy & Amindive Islands	7	7	100.0	—	—
All-India	726,111	324,123	44.6	57,825	17.8

TABLE 5A

AREA UNDER DIFFERENT TYPES OF IRRIGATION

(*thousand acres*)

State	*Area irrigated in India 1958-59*				
	Canals	*Tanks*	*Wells*	*Other sources*	*Total*
Andhra Pradesh	3,109	2,944	728	265	7,046
Assam (excluding NEFA)	899	—	—	634	1,533
Bihar	1,434	1,024	681	1,951	5,090
Maharashtra	722	485	2,484	116	3,807
Jammu & Kashmir	703	—	7	26	736
Kerala	450	79	35	315	879
Madhya Pradesh	960	281	741	90	2,072
Madras	2,075	2,076	1,346	89	5,586
Mysore	473	859	304	286	1,922
Orissa	556	1,223	94	541	2,414
Punjab	4,981	17	2,322	41	7,361
Rajasthan	818	774	1,944	35	3,571
Uttar Pradesh	4,611	1,084	5,754	688	12,137
West Bengal	1,922	910	39	468	3,339
Delhi	32	5	42	—	79
Himachal Pradesh	—	—	—	96	96
Manipur	149	—	—	—	149
Tripura	—	—	—	8	8
Total	23,894	11,761	16,521	5,649	57,825

Source: *Agricultural Situation in India*, Min. of F & A, 1962, Vol. 17, No. 8.

TABLE 6A

CROP-WISE DISTRIBUTION OF GROSS CROPPED AREA AND GROSS IRRIGATED AREA, 1958-59.

Crop	*Gross cropped area (thousand acres)*	*Gross irrigated area (thousand acres)*	*Percentage of crop area irrigated*
Foodgrains	283,165	53,029	18.7
Rice	81,000	29,765	36.7
Wheat	31,164	9,923	31.8
Barley	8,190	3,396	41.5
Maize	10,534	1,101	10.5
Jowar	44,438	1,535	3.5
Bajra	28,185	849	3.0
Ragi	6,313	1,071	17.0
Other cereals & millets	13,143	292	2.2
Gram	24,828	2,983	12.0
Other pulses	35,370	2,114	6.0
Sugarcane	4,822	3,310	68.6
Other food crops[1]	10,437	3,111	27.1
Oilseeds[2]	32,304	1,043	3.2
Cotton	19,683	2,465	12.5
Jute & mesta	2,615	(a)	
Other fibres	764	(a)	
Tobacco	967	197	20.4
Fodder Crops	13,401	2,558	19.1
Tea	784	(a)	
Coffee	295	(a)	
Rubber	286	(a)	
Other non-food crops	2,209	878 (a)	39.7 (b)
Total	372,762	66,591	17.9

[1]Includes (in '000 acres) chillies (1,458), black pepper (231), all fruits (2,457), potatoes (832) and other vegetables (2,932).

[2]Includes (in '000 acres) groundnut (15,451), castor (1,212), sesamum (4,666) rape and mustard (3,059), linseed (3,365) and coconut (1,713).

(a) Included in other non-food crops.

(b) For all other non-food crops marked (a).

Source: Various Ministry of Food & Agriculture Publications.

TABLE 7A

PERCENTAGE OF LAND RECEIVING DIFFERENT INTENSITIES OF RAINFALL

Average annual rainfall	*Percentage to the total area*
Above 75″ (above 185 cms)	11
Between 50-75″ (125—185 cms)	21
30-50″ (75—125 cms)	37
15-30″ (37—75 cms)	24
Below 15″ (below 37 cms)	7

Source: Indian Agriculture in Brief, Directorate of Economics and Statistics

TABLE 8A

ASSURED, MEDIUM AND DRY RAINFALL REGIONS

Rainfall	*Area (million acres)*
Assured rainfall region (50″ and above) (125 cms and above)	209
Medium rainfall regions (between 30″ and 50″) (75—125 cms)	432
Dry region (below 30″) (below 75 cms)	165
Total geographical area	806
Area for which agricultural statistics are available, i.e. reporting area	722

Source: Indian Agriculture in Brief, Directorate of Economics and Statistics.

GLOSSARY

Acid soil: A soil giving an acid reaction (below 7.0). A soil having a preponderance of H ions over OH ions in the soil solution.

Aggregate: A single mass or cluster of soil consisting of many soil particles held together, such as a clod, crumb or gravel.

Alkali soil: A soil that contains sufficient exchangeable sodium to interfere with the growth of most crop plants—either with or without appreciable quantities of soluble salts.

Alkaline: A chemical term referring to "basic" reaction where the pH is above 7 as distinguished from "acid" reaction where the pH is below 7.

Alluvial soil: An azonal group of soils developed from transported and relatively recently deposited material (by streams) characterised by a weak modification (or none) of the original material by soil-forming processes.

Alluvium: Fine material, such as sand, silt, clay or other sediments, deposited on land by streams.

Ammonification: Production of ammonia as a result of biological decomposition of organic nitrogen compounds.

Anion: An ion carrying a negative charge of electricity.

Arid: A climate that is characterised by low rainfall and high rate of evaporation. Arid climate is usually defined as less than 10 inches of precipitation per year and semi-arid as between 10 and 20 inches per year.

Atmosphere: The whole mass of air surrounding the earth.

Availability: State of nutrients being taken up by plants. Availability is determined by water solubility, solubility of solution that collects near plant roots, and by analysis of plants.

Azonal soils: Soils without distinct genetic horizons. A soil order.

Base exchange: The replacement of cations, held on the soil complex by other cations.

Buffering: The resistance of a substance (soils, clay) to an abrupt change in acidity or alkalinity.

Calcareous soil: Soil containing sufficient $CaCO_3$ (often with $MgCO_3$) to effervesce visibly to the naked eye when treated with hydrochloric acid. Soils with at least 1% carbonate show such effervescence.

Capability ratings: The classification of land (or soil) according to its ability to produce, and practices necessary to obtain good production economically.

Capillary Water: Water held by adhesion and surface tension forces as a film around particles and in capillary spaces. Water moves in any direction where the capillary tension is greatest.

Carbon - Nitrogen ratio: The proportion of organic carbon to organic nitrogen in soil or organic matter. Number obtained by dividing the percentage of organic carbon by the percentage of nitrogen.

Catch crop: A crop seeded with one of the regular crops in a rotation or between the growing period separating two regular crops, for the purpose of adding to soil, organic matter and nitrogen.

Catena: A sequence of soils from similar parent materials and of similar age in areas of similar climate, but characters differ because of relief and drainage.

Cation: An ion carrying a positive charge of electricity.

Cation exchange capacity: The equivalents of cations per 100 gms of soil which can be held by surface forces and which can be replaced by other cations. The term as applied to soils is synonymous with base exchange capacity, but is somewhat more precise in meaning. It is usually determined by leaching soils with neutral ammonium acetate and determining the absorbed ammonia.

Chlorosis: Yellowing of green portions of plants, particularly, the leaves. May be caused by disease organisms, unavailability of nutrients, or other factors.

Clay: The soil mineral grains or particles less than 0.002 mm in diameter.

Clay pan: A dense and heavy soil horizon underlying the upper part of the soil; hard when dry and plastic or stiff when wet, presumably found in part by clay brought in by percolating water.

Colloid: The small particles (organic or inorganic) having very small diameters in proportion to the surface area. Colloids are characterised by high base exchange, can pass through filter papers, and can be dispersed.

Cover crop: A crop planted to cover or protect the soil for a certain part of the year. May be used as green manure.

Deflocculate: To separate or break down soil aggregate into their component particles. Usually refers to clay and particles of colloidal dimensions.

Denitrification: Reduction of nitrate to nitrites and ammonia to free nitrogen or any part of this process.

Desalinization: Removal of excess soluble salts from the soil, usually by leaching.

Dispersion: The destroying of soil structure (breaking up of the granules) so that individual soil particles behave as separate units.

Erosion: The wearing away of the land surface by falling or running water, wind or other natural agents such as gravity.

Essential elements: Those elements that must be present for a plant to grow. They may be primary or major (those needed in quantity) or minor elements (those needed in small amounts).

Exchangeable ions: Those ions held on the soil complex that may be replaced by other ions. There are ions held so tightly that they cannot be exchanged.

Exchangeable sodium percentage: This term indicates the degree of saturation of the soil exchange complex with sodium and is determined as follows:

$$\text{ESP} = \frac{\text{Exch. Sodium (m.e. per 100 gms soil)}}{\text{Cation Exch. capcity (m.e. per 100 gms soil)}} \times 100$$

Family, soil: A category in soil classification between series and great soil group, a toxonomic group of soils having similar profiles composed of one or more distinct soil series.

Field capacity: The amount of water held in the soil after the excess of gravitational (free) water is drained away.

Fertility: The quality of soil to aid plant-growth by supplying nutrients in desirable proportions and amounts.

Fertilizer Elements: The three primary essential elements, nitrogen, phosphorus and potassium. They are so named because they are artificially supplied to the soil in manures or commercial fertilizers.

Flocculate: To aggregate individual particles into small groups or clusters as of clay particles.

Granulation: The cementation of particles into masses as grains, aggregates or clumps, essentially a result of flocculation and aggregation.

Gravitational Water: The water that moves under the force of gravity; it is not retained in the soil.

Great Soil Group (Soil classification): A group of soils having common internal soil characteristics, includes one or more families of soils.

Green manure: Green material, usually in the form of a crop grown on the land, incorporated in the soil to increase organic matter and fertility.

Hard pan: Hardened or cemented soil horizon. Clay pan may or may not be hard pan and should be differentiated.

Horizon, soil: A layer of soil approximately parallel to the land surface with more or less well-defined characteristics produced through the operation of soil building processes. Each layer differs from the one above or below in some characteristic fashion.

Humus: Well-decomposed, more or less, stable part of organic matter of the soil, which has lost its original structure and that has not yet been reduced to the simple end products.

Hygroscopic water: Water which is absorbed from atmospheric water vapour and held on the surface of particles by forces of adhesion.

Immature soil: A young soil. A soil lacking a well-developed profile.

Indicator plants: Plants which are characterised by and reflect specific growing condition either by presence or character of growth.

Indurated: Cemented, hardened or rock-like, as a true hard pan, which will not soften when wetted.

Infiltration: The process by which water enters the soil through the surface. The rate at which it soaks into the soil is called "infiltration capacity."

Intrazonal soil: Any of the great group of soils with more or less well-developed soil characteristics that reflect the dominating influence of some local factor of relief, parent material or age, over the normal effect of the climate and vegetation.

Ion: An electrically charged particle, element or group of elements.

Irrigation: Artificial application of water to soil for the purpose of supplying moisture essential to plant growth.

Land reclamation: Making land capable of more intensive use by changing its character or environment or both through operations requiring collective effect. Does not include clearing of land from stumps, stones, etc. or simple techinque of erosion control that can be effected by the individual.

Leaching: The process of removal of soluble material from soil by the passage of water through it. This is a primary step in the improvement of saline soils.

Lithosols: An azonal group of soils having no clearly defined morphology and consisting of a mass of rock fragments from consolidated rocks which are imperfectly weathered. Found primarily on sloping ground.

Marl: A soft earthy deposit consisting chiefly of $CaCO_3$ mixed with sand, clay, organic matter and other impurities in varying proportions. Frequently used as a liming material.

Mature soil: A soil with well-developed characteristics, produced by the natural processes of soil formation and in equilibrium with its environment.

Mechanical analysis: The separation by mechanical means of the different size groups (separates) and the determination of the percentage of each group in a given soil sample.

Metamorphic rock: A rock the constitution of which has undergone pronounced alteration. Such changes are generally effected by the combined action of pressure, heat, and water, frequently resulting in a more compact and more highly crystalline condition of the rock. Gneiss, schist, and marble are common examples.

Mineralization: The conversion of an element that is in organic combination to the available form as a result of microbial decomposition.

Morphology, soil: The physical constitution of the soil, including the texture, structure, porosity, consistence and colour of the various soil horizons, their thickness, and their arrangement in the soil profile.

Neutral soil: A soil that is not significantly acid or alkaline; strictly, one having a pH of 7.0.

Nitrification: Formation of nitrates from ammonia as in soils by soil organisms.

Nitrogen fixation: Conversion of free nitrogen into nitrogen compounds by symbiotic or non-symbiotic nitrogen activity.

Parent material: The slightly altered or unweathered material beneath the solum; similar to that from which the soil was formed.

Parent rock: The rock from which parent materials of soil are formed.

Permeability: The specific property of a soil which is a measure of the readiness with which the soil transmits water. The permeability is a velocity, and for agricultural purposes it can be conveniently expressed either in inches per hour or centimetres per hour. For most soils both the chemical nature of the water used and the history and treatment of the sample greatly affect the permeability.

pH: A notation to designate or indicate the degree of acidity or a alkalinity of systems. Technically, the common logarithm of the reciprocal of the hydrogen ion concentration (gms per litre) of a system.

Phase, soil: That part of a soil type having minor variations in characteristics used in soil classification from the characteristics normal for the type, although they may be of great practical importance. The variations are chiefly in such external characteristics as relief, stoniness, or erosion.

Porosity: The fraction of the soil volume not occupied by the soil particles. In other words, porosity is the ratio of the sum of the volumes of the solid, liquid and gaseous phases of the soil.

Productivity (of soil): The capability of a soil for producing a specified plant or sequence of plants under a specified system of management.

Profile, soil: A vertical cross section of the soil from the surface into the underlying unweathered material.

Puddling: The process by which a soil loses granular structure and becomes deflocculated. It is caused by excessive water, excessive handling or tilling or deflocculating agents.

Reaction, soil: The degree of acidity or alkalinity of the soil mass expressed in pH values or in words.

Reclamation: Conversion of poor quality land or other resources to more useful resources by special practices.

Relief: The elevations or inequalities of a land surface considered collectively.

Reversion: Process by which plant nutrients become unavailable because of physical and chemical tie-up.

Saline soil: A non-alkali soil containing sufficient soluble salts to impair its productivity.

Sand: Small rock or mineral fragments having diameters ranging from 2 to 0.2 mm in the case of coarse sand, and 0.2 to 0.02 mm in the case of fine sand.

Sedimentary rock: A rock composed of particles deposited from suspension in water.

Series, soil: A group of soils having genetic horizons similar in differentiating characteristics and arrangement in the soil profile, except for the texture of the surface soil, and developed from a particular type of parent material. A series may include two or more soil types differing from one another in the texture of the surface soils.

Soil: The natural medium for the growth of land plants on the surface of the earth. A natural body on the surface of the earth in which plants grow; composed of organic and mineral materials.

Soil Morphology: The physical constitution of the soil including the texture, structure, porosity, consistence and colour of the various soil horizons, their thickness and their arrangement in the soil profile.

Soil type: A group of soils having genetic horizons similar in differentiating characteristics, including texture and arrangement in the soil profile, and developed from a particular type of parent material.

Solum: The upper part of the soil profile, above the parent material, in which the processes of soil formation are taking place. In mature soils this includes the A-and B-horizons.

Stratified: Composed of, or arranged in, strata or layers, as stratified alluvium. The term is applied to geological materials. Those layers in soils that are produced by the processes of soil formation are called horizons; those inherited from the parent materials are called strata.

Strip Cropping: Strip cropping is the practice of growing ordinary farm crops in long strips of variable widths, across the line of slope, approximately on the contour, in which densely-growing crops are seeded in alternate strips with cleantilled crops.

Structure, soil: The morphological aggregates in which the individual soil particles are arranged.

Subsoil: Roughly, that part of the solum below plow depth.

Surface soil: That part of the upper soil of arable soils commonly stirred by tillage implements or an equivalent (5 to 8 inches) in non-arable soils.

Terrace, (for control of run-off, soil erosion, or both): A broad-surface channel or embankment constructed across the sloping lands, on or approximately on, contour lines, at specific intervals. The terrace intercepts surplus run-off to retard it for infiltration or to direct the flow to an outlet at non-erosive velocity.

Texture, soil: The relative proportion of the various size groups of individual soil grains indicates the coarseness or fineness of the soil.

Tilth, soil: The physical condition of the soil in relation to plant growth. A term indicating the conditions of soil structure produced by tillage or cultivation.

Top soil: A general term applied to the surface portion of the soil, including the average plow depth (surface soil) or the A-horizon, where this is deeper than the plow depth.

Transpiration Ratio (Water Requirement): Ratio of weight of water absorbed and transpired by the plant to the weight of the dry matter produced. (Ratios vary from about 250 to 1000 pounds of water per pound of dry matter.)

Vesicular structure: Soil structure characterized by round or egg-shaped cavities or vesicles.

Water table: The upper limit of the part of the soil or underlying material wholly saturated with water.

Weathering: The physical and chemical disintegration or decomposition of rocks and minerals under natural conditions.

Wilting coefficient (wilting point): The percentage of water in the soil (based on dry weight of the soil) when permanent wilting of plants occurs. It refers to that moisture content at which soil cannot supply water at a rate sufficient to maintain the turgor of a plant and it permanently wilts.

Zonal soils: Any one of the great groups of soils having well-developed soil characteristics that reflect the influence of the active factors of soil genesis, climate and living organisms, chiefly vegetation.

INDEX